2ND EDITION

Trees *of* Pennsylvania

Field Guide

Stan Tekiela

Adventure Publications
Cambridge, Minnesota

Edited by Sandy Livoti and Dan Downing

Cover, book design and leaf illustrations by Jonathan Norberg

Tree illustrations by Julie Martinez

Cover photo: Quaking Aspen by Stan Tekiela
All photos copyright by Stan Tekiela unless otherwise noted.

Steven J. Baskauf: 144 (fruit); **Rick & Nora Bowers:** 74 (Immature fruit), 80 (flower), 162 (main), 166 (flower); **Dominique Braud/DPA*:** 90 (fruit); **E. R. Degginger/DPA*:** 108 (fruit); **Dembinsky Photo Associates:** 246 (bark); **Dudley Edmondson:** 74 (bark, flower, mature fruit & main), 90 (flower), 112 (main), 120 (flower & fruit), 144 (bark & main), 162 (bark), 166 (bark, fruit & main), 180 (fruit), 188 (bark, flower & main), 192 (bark & main), 212 (flower), 232 (flower); **Michael P. Gadomski/DPA*:** 238 (main); **H. Garrett/DPA*:** 180 (flower); **Jerry & Barbara Jividen:** 238 (bark); **Ted Nelson/DPA*:** 246 (flower); and **David Wisse/DPA*:** 246 (fruit).

*Dembinsky Photo Associates

The following Bugwood (weedimages.org) images below are licensed under the Creative Commons Attribution 3.0 (CC BY 3.0 US) License, available here: https://creativecommons.org/licenses/by/3.0/deed.en.
Franklin Bonner/USFS: 148 (fruit), 200 (fruit); **Bill Cook/Michigan State University:** 158 (fruit), 174 (flower); **Dow Gardens:** 184 (fruit); **Keith Kanoti/Maine Forest Service:** 172 (flower); **Pennsylvania Department of Conservation and Natural Resources, Forestry:** 188 (fruit); **Rob Routledge/Sault College:** 184 (flower).

Images used under license from Shutterstock.com:
John De Winter: 96 (flower); **eleonimages:** 186 (flower); **islavicek:** 192 (fruit); **J Need:** 164 (flower); **Nikolay Kurzenko:** 168 (flower), 190 (flower); **lightrain:** 158 (flower); **Bruce MacQueen:** 122 (flower); **simona pavan:** 228 (flower); **Puffin's Pictures:** 98 (flower); **Jordi Roy:** 94 (fruit), 186 (fruit); **sam-bhatharu:** 130 (fruit); **Shigeyoshi Umezu:** 110 (fruit); and **Iva Villi:** 262 (flower).

10 9 8 7 6 5 4 3 2
Trees of Pennsylvania Field Guide
First Edition 2004
Second Edition 2021
Copyright © 2004 and 2021 by Stan Tekiela
Published by Adventure Publications
An imprint of AdventureKEEN
310 Garfield Street South
Cambridge, Minnesota 55008
(800) 678-7006
www.adventurepublications.net
All rights reserved
Printed in China
ISBN 978-1-64755-204-6 (pbk.), ISBN 978-1-64755-205-3 (ebook)

TABLE OF CONTENTS

PENNSYLVANIA AND TREES

Pennsylvania is a great place for anyone interested in trees. With *Trees of Pennsylvania Field Guide*, you'll be able to quickly identify 117 of the most common trees in Pennsylvania—most of which are native to the state. This guide also includes a number of common non-native trees that have been naturalized in Pennsylvania. This book makes no attempt to identify cultivated or nursery trees.

Because this book is a unique all-photographic field guide just for Pennsylvania, you won't have to page through photographs of trees that don't grow in the state, or attempt to identify live trees by studying black-and-white line drawings.

WHAT IS A TREE?

For the purposes of this book, a tree is defined as a large woody perennial plant, usually with a single erect trunk, standing at least 15 feet (4.5 m) tall, with a well-defined crown. *Trees of Pennsylvania Field Guide* helps you observe some basic characteristics of trees so you can identify different species confidently.

HOW THIS BOOK IS ORGANIZED

To identify a tree, you'll want to start by looking at the thumb tabs in the upper right-hand corner of the text pages. These tabs define the sections of the book. They combine several identifying features of a tree (main category, needle or leaf type and attachment) into one icon.

It's possible to identify trees using this field guide without learning about categories, leaf types and attachments. Simply flip through the pages to match your sample to the features depicted on the thumb tabs. Once you find the correct section, use the photos to find your tree. Or, you may want to learn more about the features of trees in a methodical way, using the following steps to narrow your choices to just a few photos.

1. First, determine the appropriate section and find the right icon by asking these questions: Is the tree coniferous or deciduous? If it is a conifer, are the needles single, clustered or scaly? If it is deciduous, is the leaf type simple, lobed or compound, and do leaves attach to twigs in an opposite or alternate pattern?

2. Next, simply browse through the photos in that section to find your tree. Or, to further narrow your choices, use the icon in the lower right-hand corner of the text pages. These icons are grouped by the general shape of the needle or leaf, and they increase in size as the average size of the needle or leaf increases.

3. Finally, by examining the full-page photos of needles or leaves, studying the inset photos of bark, flowers, fruit or other special features and considering information on text pages, you should be able to confidently identify the tree.

While these steps briefly summarize how you can use this book, it is quite helpful to learn more about how the sections are grouped by reading the Identification Step-by-Step section.

IDENTIFICATION STEP-BY-STEP
Conifer or Deciduous

Trees in this field guide are first grouped into two main categories that consist of 19 conifers and 98 deciduous trees.

Trees with evergreen needles that remain on branches year-round and have seeds in cones are conifers. Some examples of these are pines and spruces. The exceptions in this main category are the Bald Cypress and the Tamarack, conifers that behave like deciduous trees, shedding their needles in autumn. Trees with broad flat leaves that fall off their branches each autumn are deciduous. Some examples of these are oaks and maples.

You will see by looking at the thumb tabs that trees with needles (conifers) are shown in the first sections of the book, followed by trees with leaves (deciduous).

Needle or Leaf Type
CONIFER GROUP:
Single, Clustered or Scaly Needles

SINGLE

CLUSTERED
(Range of
2–30 needles)

SCALY

If the tree is a conifer, the next step is to distinguish among single, clustered and scaly needles. Begin by checking the number of needles that arise from one point. If you see only one needle arising from one point, look in the single needle section. Conifers with single needles are shown first. If there are at least two needles arising from one point, turn to the clustered needles section. This second section is organized by the number of needles in a cluster. If you are trying to identify needles that overlap each other and have a scale-like appearance, unlike the other needles, you will find this type in the scaly needles section.

DECIDUOUS GROUP:
Simple, Lobed or Compound

SIMPLE

LOBED

COMPOUND

TWICE COMPOUND

PALMATE COMPOUND

If the tree is deciduous, the next step is to determine the leaf type. Many of the simple leaves have a basic shape such as oval, round or triangular. Other simple leaves are lobed, identified by noticeable indentations along their edges. Simple leaves without lobes are grouped first, followed by the lobed leaf groups.

If a leaf is composed of smaller leaflets growing along a single stalk, you'll find this type in the compound leaf sections. When a leaf has small leaflets growing along the edge of a thinner secondary stalk, which is in turn attached to a thicker main stalk, check the twice compound section. If the leaf has leaflets emerging from a common central point at the end of a leafstalk, look in the palmate compound section.

Leaf Attachment

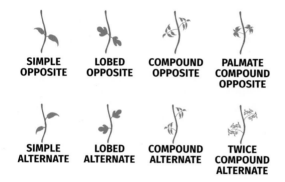

| SIMPLE OPPOSITE | LOBED OPPOSITE | COMPOUND OPPOSITE | PALMATE COMPOUND OPPOSITE |
| SIMPLE ALTERNATE | LOBED ALTERNATE | COMPOUND ALTERNATE | TWICE COMPOUND ALTERNATE |

For deciduous trees, once you have determined the appropriate leaf type, give special attention to the pattern in which the leaves are attached to the twig. Trees with leaves that attach directly opposite of each other on a twig are grouped first in each section, followed by trees with leaves that attach alternately. The thumb tabs are labeled "opposite" or "alternate" to reflect the attachment group. All the above features (main category, needle or leaf type and attachment) are depicted in one icon for easy use.

Needle or Leaf Size

Once you have found the correct section by using the thumb tabs, note that the section is further loosely organized by needle or leaf size from small to large. Size is depicted in the needle or leaf icon located in the lower right-hand corner of text pages. This icon also reflects the shape of the needle or leaf. For example, the icon for the Amur Maple, which has a leaf size of 2–4 inches (5–10 cm), is smaller than the icon for the Norway Maple with a leaf size of 5–7 inches (12.5–18 cm). Measurement of any deciduous leaf extends from the base of the leaf (excluding the leafstalk) to the tip.

Using Photos and Icons to Confirm the Identity

After using the thumb tabs to narrow your choices, the last step is to confirm the tree's identity. First, compare the full-page photo of the leaves and twigs to be sure they look similar. Next, study the color and texture of the bark, and compare it to the inset photo. Then consider the information given about the habitat and range.

Sometimes, however, it is a special characteristic, such as flowers, fruit or thorns (described and/or pictured), that is an even better indicator of the identity. In general, if it's spring, check for flowers. During summer, look for fruit. In autumn, note the fall color.

Another icon is also included for each species to show the overall shape of the average mature tree and how its height compares with a two-story house. For trees with an average height of more than 50 feet (15 m), this icon is shown on a slightly smaller scale.

STAN'S NOTES

Stan's Notes is fun and fact-filled with many gee-whiz tidbits of interesting information, such as historical uses, other common names and much more. Most information given in this descriptive section cannot be found in other tree field guides.

CAUTION

In Stan's Notes, it's occasionally mentioned that parts of some trees were used for medicine or food. While some find this interesting, DO NOT use this field guide to identify edible or medicinal trees. Certain trees in the state have toxic properties or poisonous look-alikes that can cause severe problems. Do not take the chance of making a mistake. Please enjoy the trees of Pennsylvania with your eyes, nose or with your camera. In addition, please don't pull off leaves, cut branches or attempt to transplant any trees. Nearly all of the trees you will see are available at your local garden centers. These trees have been cultivated and have not been uprooted from the wild. Trees are an important part of our natural environment, and leaving a healthy tree unharmed will do a great deal to help keep Pennsylvania the wondrous place it is.

Enjoy the Trees!

LEAF BASICS

It's easier to identify trees and communicate about them when you know the names of the different parts of a leaf. For instance, it is more effective to use the word "sinus" to indicate an indentation on an edge of a leaf than to try to describe it.

The following illustrations show coniferous needles in cross section and the basic parts of deciduous leaves. The simple/lobed leaf and compound leaf illustrations are composites of leaves and should not be confused with any actual leaf of a real tree.

Needle Cross Sections

square   flat   triangular   round

Simple/Lobed Leaf

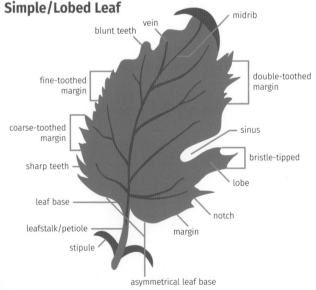

midrib

vein

blunt teeth

fine-toothed margin

double-toothed margin

coarse-toothed margin

sinus

bristle-tipped

sharp teeth

lobe

leaf base

notch

leafstalk/petiole

margin

stipule

asymmetrical leaf base

Compound Leaf

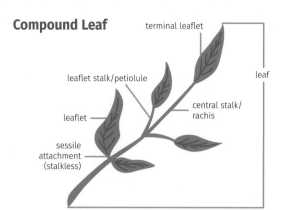

terminal leaflet

leaflet stalk/petiolule

leaflet

sessile attachment (stalkless)

central stalk/ rachis

leaf

FINDING YOUR TREE IN A SECTION

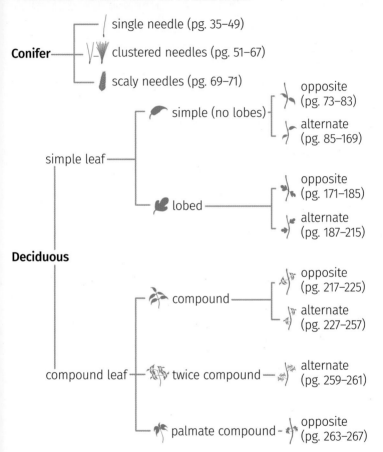

Conifer — single needle (pg. 35–49)
clustered needles (pg. 51–67)
scaly needles (pg. 69–71)

Deciduous

simple leaf — simple (no lobes) — opposite (pg. 73–83)
alternate (pg. 85–169)

lobed — opposite (pg. 171–185)
alternate (pg. 187–215)

compound leaf — compound — opposite (pg. 217–225)
alternate (pg. 227–257)

twice compound — alternate (pg. 259–261)

palmate compound — opposite (pg. 263–267)

The smaller needles and leaves tend to be toward the front of each section, while larger sizes can be found toward the back. Check the icon in the lower right corner of text pages to compare relative size.

SILHOUETTE QUICK COMPARES

To quickly narrow down which mature tree you've found, compare its rough outline with the samples found here. For a sense of scale, we've included the tree's height range compared with a drawing of a typical U.S. house. Obviously, tree heights and general shapes can vary significantly across individuals, but this should help you rule out some possible options, hopefully pointing you in the right direction. Once you've found a possible match, turn to the specified page and confirm or rule it out by examining the photos of bark and leaves and the accompanying text.

Common Prickly-ash
5–15'
pg. 235

Poison Sumac
5–20'
pg. 247

American Plum
10–15'
pg. 117

Wild Apple
10–15'
pg. 109

Buttonbush
10–20'
pg. 81

Crab Apple
10–20'
pg. 107

European Buckthorn
10–20'
pg. 73

Juneberry
10–20'
pg. 139

Nannyberry
10–20'
pg. 79

Pussy Willow
10–20'
pg. 113

Russian-olive
10–20'
pg. 133

Smooth Sumac
10–20'
pg. 249

Staghorn Sumac
10–20'
pg. 251

Gray Birch
10–30'
pg. 101

Pin Cherry
10–30'
pg. 119

Striped Maple
10–30'
pg. 185

Amur Maple
15–20'
pg. 171

Blue Beech
15–25'
pg. 143

Eastern Redbud
15–25'
pg. 155

European Mountain-ash
15–25'
pg. 231

Hawthorn
15–25'
pg. 137

American Mountain-ash
15–30'
pg. 233

Choke Cherry
15–35'
pg. 121

American Bladdernut
20–25'
pg. 219

Common Hoptree
20–25'
pg. 227

Eastern Wahoo
20–25'
pg. 77

Eastern Flowering Dogwood
20–30'
pg. 75

Mountain Maple
20–30'
pg. 173

Pawpaw
20–30'
pg. 167

Red Mulberry
20–30'
pg. 151

Witch-hazel
20–30'
pg. 159

Ironwood
20–40'
pg. 141

Ohio Buckeye
20–40'
pg. 265

Table Mountain Pine
20–40'
pg. 51

Alternate-leaf Dogwood
25–35'
pg. 147

Blackjack Oak
25–40'
pg. 197

Black Spruce
25–50'
pg. 39

Eastern Redcedar
25–50'
pg. 69

Common Persimmon
30–40'
pg. 157

Osage-orange
30–40'
pg. 153

Black Locust
30–50'
pg. 253

Boxelder
30–50'
pg. 217

Eastern Whitecedar
30–50'
pg. 71

Siberian Elm
30–50'
pg. 85

Post Oak
30–60'
pg. 201

Sassafras
30–60'
pg. 189

Virginia Pine
30–60'
pg. 53

Scotch Pine
30–80'
pg. 55

Black Ash
40–50'
pg. 221

Austrian Pine
40–60'
pg. 57

Black Oak
40–60'
pg. 211

Black Willow
40–60'
pg. 115

Butternut
40–60'
pg. 257

Colorado Spruce
40–60'
pg. 43

Cucumbertree
40–60'
pg. 165

Eastern Hemlock
40–60'
pg. 47

Ginkgo
40–60'
pg. 131

Hackberry
40–60'
pg. 135

Honey Locust
40–60'
pg. 259

Horse-chestnut
40–60'
pg. 267

Kentucky Coffeetree
40–60'
pg. 261

Norway Maple
40–60'
pg. 183

Paper Birch
40–60'
pg. 103

Red Maple
40–60'
pg. 175

Shagbark Hickory
40–60'
pg. 241

Swamp White Oak
40–60'
pg. 207

SILHOUETTE QUICK COMPARES, *continued*

White Ash
40–60'
pg. 223

White Poplar
40–60'
pg. 187

White Spruce
40–60'
pg. 37

Black Maple
40–70'
pg. 179

Quaking Aspen
40–70'
pg. 91

Tamarack
40–70'
pg. 67

Mockernut Hickory
40–80'
pg. 243

Red Pine
40–80'
pg. 59

Weeping Willow
40–80'
pg. 111

Green Ash
50–60'
pg. 225

Shingle Oak
50–60'
pg. 125

American Basswood
50–70'
pg. 161

Balsam Poplar
50–70'
pg. 97

Bigtooth Aspen
50–70'
pg. 93

Black Tupelo
50–70'
pg. 149

Chinquapin Oak
50–70'
pg. 127

Norway Spruce
50–70'
pg. 45

Pin Oak
50–70'
pg. 199

Pitch Pine
50–70'
pg. 63

Red Oak
50–70'
pg. 213

Red Spruce
50–70'
pg. 41

Slippery Elm
50–70'
pg. 89

Sugar Maple
50–70'
pg. 177

Tree-of-heaven
50–70'
pg. 229

White Oak
50–70'
pg. 209

Yellow Birch
50–70'
pg. 105

Yellow Buckeye
50–70'
pg. 263

Balsam Fir
50–75'
pg. 49

Black Cherry
50–75'
pg. 123

Black Walnut
50–75'
pg. 255

Northern Catalpa
50–75'
pg. 83

Bur Oak
50–80'
pg. 215

Bitternut Hickory
50–100'
pg. 237

Scarlet Oak
60–70'
pg. 203

American Beech
60–80'
pg. 145

English Oak
60–80'
pg. 195

Lombardy Poplar
60–80'
pg. 95

Red Hickory
60–80'
pg. 239

American Chestnut
60–90'
pg. 169

Chestnut Oak
60–90'
pg. 129

Shumard Oak
60–90'
pg. 205

Sycamore
60–90'
pg. 163

Shellbark Hickory
70–90'
pg. 245

American Elm
70–100'
pg. 87

Eastern Cottonwood
70–100'
pg. 99

Eastern White Pine
70–100'
pg. 65

Shortleaf Pine
70–100'
pg. 61

Silver Maple
75–100'
pg. 181

SILHOUETTE QUICK COMPARES, *continued*

Bald Cypress
80–100'
pg. 35

Sweetgum
80–100'
pg. 191

Tulip-tree
80–100'
pg. 193

NEEDLE AND LEAF QUICK COMPARES

To help you differentiate among similar-looking tree species, compare your finds with the following leaf images. For each species, we've also included information about the leaf shape and attachment, which can help quickly point you in the right direction.

Note: Leaf images are not to scale.

Bald Cypress
single needle
pg. 35

White Spruce
single needle
pg. 37

Black Spruce
single needle
pg. 39

Red Spruce
single needle
pg. 41

Colorado Spruce
single needle
pg. 43

Norway Spruce
single needle
pg. 45

Eastern Hemlock
single needle
pg. 47

Balsam Fir
single needle
pg. 49

Table Mountain Pine
single needle
pg. 51

Virginia Pine
clustered needles
pg. 53

Scotch Pine
clustered needles
pg. 55

Austrian Pine
clustered needles
pg. 57

Red Pine
clustered needles
pg. 59

Shortleaf Pine
clustered needles
pg. 61

Pitch Pine
clustered needles
pg. 63

Eastern White Pine
clustered needles
pg. 65

Tamarack
clustered needles
pg. 67

Eastern Redcedar
scaly needles
pg. 69

Eastern Whitecedar
scaly needles
pg. 71

European Buckthorn
simple opposite
pg. 73

Eastern Flowering Dogwood
simple opposite
pg. 75

Eastern Wahoo
simple opposite
pg. 77

Nannyberry
simple opposite
pg. 79

Buttonbush
simple opposite
pg. 81

Northern Catalpa
simple opposite
pg. 83

Siberian Elm
simple alternate
pg. 85

American Elm
simple alternate
pg. 87

Slippery Elm
simple alternate
pg. 89

Quaking Aspen
simple alternate
pg. 91

Bigtooth Aspen
simple alternate
pg. 93

Lombardy Poplar
simple alternate
pg. 95

Balsam Poplar
simple alternate
pg. 97

Eastern Cottonwood
simple alternate
pg. 99

Gray Birch
simple alternate
pg. 101

Paper Birch
simple alternate
pg. 103

Yellow Birch
simple alternate
pg. 105

Crab Apple
simple alternate
pg. 107

Wild Apple
simple alternate
pg. 109

Weeping Willow
simple alternate
pg. 111

Pussy Willow
simple alternate
pg. 113

Black Willow
simple alternate
pg. 115

American Plum
simple alternate
pg. 117

Pin Cherry
simple alternate
pg. 119

Choke Cherry
simple alternate
pg. 121

Black Cherry
simple alternate
pg. 123

24

Shingle Oak
simple alternate
pg. 125

Chinquapin Oak
simple alternate
pg. 127

Chestnut Oak
simple alternate
pg. 129

Ginkgo
simple alternate
pg. 131

Russian-olive
simple alternate
pg. 133

Hackberry
simple alternate
pg. 135

Hawthorn
simple alternate
pg. 137

Juneberry
simple alternate
pg. 139

Ironwood
simple alternate
pg. 141

Blue Beech
simple alternate
pg. 143

American Beech
simple alternate
pg. 145

Alternate-leaf Dogwood
simple alternate
pg. 147

Black Tupelo
simple alternate
pg. 149

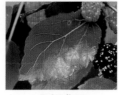
Red Mulberry
simple alternate
pg. 151

Osage-orange
simple alternate
pg. 153

Eastern Redbud
simple alternate
pg. 155

Common Persimmon
simple alternate
pg. 157

Witch-hazel
simple alternate
pg. 159

American Basswood
simple alternate
pg. 161

Sycamore
simple alternate
pg. 163

Cucumbertree
simple alternate
pg. 165

Pawpaw
simple alternate
pg. 167

American Chestnut
simple alternate
pg. 169

Amur Maple
lobed opposite
pg. 171

Mountain Maple
lobed opposite
pg. 173

Red Maple
lobed opposite
pg. 175

Sugar Maple
lobed opposite
pg. 177

Black Maple
lobed opposite
pg. 179

Silver Maple
lobed opposite
pg. 181

Norway Maple
lobed opposite
pg. 183

Striped Maple
lobed opposite
pg. 185

White Poplar
lobed alternate
pg. 187

Sassafras
lobed alternate
pg. 189

Sweetgum
lobed alternate
pg. 191

Tulip-tree
lobed alternate
pg. 193

English Oak
lobed alternate
pg. 195

Blackjack Oak
lobed alternate
pg. 197

Pin Oak
lobed alternate
pg. 199

Post Oak
lobed alternate
pg. 201

Scarlet Oak
lobed alternate
pg. 203

Shumard Oak
lobed alternate
pg. 205

Swamp White Oak
lobed alternate
pg. 207

White Oak
lobed alternate
pg. 209

Black Oak
lobed alternate
pg. 211

Red Oak
lobed alternate
pg. 213

Bur Oak
lobed alternate
pg. 215

Boxelder
compound opposite
pg. 217

American Bladdernut
compound opposite
pg. 219

Black Ash
compound opposite
pg. 221

White Ash
compound opposite
pg. 223

Green Ash
compound opposite
pg. 225

Common Hoptree
compound alternate
pg. 227

Tree-of-heaven
compound alternate
pg. 229

European Mountain-ash
compound alternate
pg. 231

American Mountain-ash
compound alternate
pg. 233

Common Prickly-ash
compound alternate
pg. 235

Bitternut Hickory
compound alternate
pg. 237

Red Hickory
compound alternate
pg. 239

Shagbark Hickory
compound alternate
pg. 241

Mockernut Hickory
compound alternate
pg. 243

Shellbark Hickory
compound alternate
pg. 245

Poison Sumac
compound alternate
pg. 247

Smooth Sumac
compound alternate
pg. 249

Staghorn Sumac
compound alternate
pg. 251

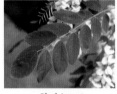

Black Locust
compound alternate
pg. 253

Black Walnut
compound alternate
pg. 255

Butternut
compound alternate
pg. 257

Honey Locust
twice compound
alternate
pg. 259

Kentucky Coffeetree
twice compound
alternate
pg. 261

Yellow Buckeye
palmate compound
opposite
pg. 263

Ohio Buckeye
palmate compound
opposite
pg. 265

Horse-chestnut
palmate compound
opposite
pg. 267

mature fruit

bark

flower

immature fruit

Common Name
Scientific name

Family: common family name (scientific family name)

Height: average range in feet and meters of the mature tree from ground to top of crown

Tree: overall description; may include a shape, type of trunk, branches or crown

Leaf/Needle: type of leaf or needle, shape, size, and attachment; may include lobes, leaflets, margin, veins, color or leafstalk

Bark: color and texture of the trunk; may include inner bark or thorns

Flower: catkin, flower; may include shape, size or color

Fruit/Cone: seed, nut, berry; may include shape, size, or color

Fall Color: color(s) that deciduous leaves turn to in autumn

Origin/Age: native or non-native to the state; average life span

Habitat: type of soil, places found, sun or shade tolerance

Range: throughout or part of Pennsylvania where the tree is found; may include places where planted

Stan's Notes: Helpful identification information, history, origin and other interesting gee-whiz nature facts.

Shape of an individual needle, needle cluster, or leaf. Use this icon to compare relative size among similarly shaped leaves.

bark

cone

knees

Bald Cypress
Taxodium distichum

Family: Cypress (Cupressaceae)

Height: 80–100' (24.5–30.5 m)

Tree: large conical tree, enlarged straight trunk with a flared base (buttress), spreading into ridges, widely spreading branches, crown often pointed

Needle: single needle, ½–¾" (1–2 cm) long, in 2 rows on slender green twigs, pointed at the tip, soft and flexible to touch, appearing feather-like, yellowish green above, whitish below

Bark: brown to gray, with narrow fibrous ridges, peeling off in long strips

Cone: green, turning gray to brown when mature, ¾–1" (2–2.5 cm) wide, solitary or in small clusters at the end of branch, several 4-sided woody cone scales

Fall Color: brown

Origin/Age: native; 500–750 years

Habitat: wet soils, swamps, by slow rivers that flood often, can grow in dry upland soils, sun to partial shade

Range: throughout, an ornamental in parks and yards

Stan's Notes: Called "Bald" since it's a deciduous conifer, losing its leaves (needles) in fall and growing new ones in spring. Produces a large flaring or fluted base, which helps stabilize it when growing in soft, wet soils. Produces large aboveground or water growths called knees (see inset). A long-lived tree, some are more than 2,000 years old and are among the oldest living things in North America. Often called the Sequoia of the East, reaching over 100 feet (30.5 m) tall and nearly 40 feet (12 m) around at the base. Decay- and insect-resistant wood has been used to build boats and bridges. Its seeds are an important food for wildlife such as ducks and deer.

bark

cone

close-up

White Spruce
Picea glauca

Family: Pine (Pinaceae)

Height: 40–60' (12–18 m)

Tree: single straight trunk, many horizontal branches sometimes sloping down, ragged conical crown

Needle: single needle, ⅓–¾" (.8–2 cm) long, stiff, pointed, square in cross section, aromatic when crushed, bluish green with a line of white dots on all sides

Bark: light gray in color, many flaky scales, inner bark is salmon pink

Cone: green, turning brown at maturity, smooth to the touch, 1–2½" (2.5–6 cm) long, single or in clusters, hanging from branch

Origin/Age: non-native; 175–200 years

Habitat: variety of soils, often grows on banks of lakes and streams, sometimes in pure stands, sun to partial shade

Range: thoughout, planted in parks, yards and along streets

Stan's Notes: Also known as Skunk Spruce because its crushed needles give off a strong odor that reminds some of skunk. Needles have a whitish cast, giving this tree its common name. Like all other species of spruce, White Spruce needles are square in cross section. Needles often last 7–10 years before falling off, leaving a raised base on the twig. Susceptible to fire and Spruce Budworm, a caterpillar that eats new needles. Lower branches die and fall off, leaving the trunk straight and lacking branches. A variety, Black Hills White Spruce (*P. glauca* var. *densata*), is a widely planted urban tree.

bark

cone

Black Spruce
Picea mariana

Family: Pine (Pinaceae)

Height: 25–50' (7.5–15 m)

Tree: small to medium-sized slender tree with a narrow pyramid shape, many dead lower branches, upper branches widely spread and drooping

Needle: single needle, ¼–1" (.6–2.5 cm) long, densely set along twig, straight, blunt-tipped, square in cross section, dull blue-green

Bark: reddish brown in color, large scales, inner bark is olive green

Cone: lavender to purple, turning brown when mature, egg-shaped (ovate), ½–1½" (1–4 cm) long, hanging from the branch

Origin/Age: non-native; 150–200 years

Habitat: wet or poorly drained soils, bogs, peats, often in pure stands or with Tamarack (pg. 67)

Range: throughout

Stan's Notes: A relatively slow-growing, long-lived tree, it is one of seven spruce species native to North America. Common along bogs and marshes, with cones usually occurring at the top of the tree. Cones mature in autumn but often don't open, remaining on the tree for up to 15 years. Heat from fire opens the cones, after which many seeds are released. This species can live as long as 200 years, but obtains a height of only 50 feet (15 m). Young twigs have tiny orange-to-brown hairs. Treetops are commonly used in planters in winter for decoration. Long fibers in the wood make it desirable for making paper. A golden-colored pitch that collects on wounds was once gathered and sold as spruce gum. Differentiated from Red Spruce (pg. 41) by its smaller cones and very different habitat.

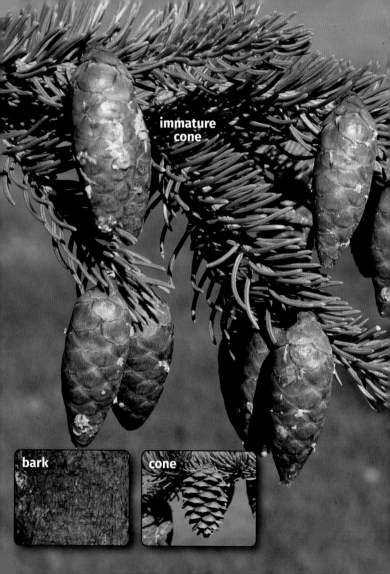

immature
cone

bark

cone

Red Spruce
Picea rubens

Family: Pine (Pinaceae)

Height: 50–70' (15–21 m)

Tree: medium tree, single straight trunk, broadest at the bottom, narrow pointed top

Needle: single needle, ½–¾" (1–2 cm) long, stout, stiff, pointed tip, 4-sided (square in cross section), shiny yellow-green with white stripes

Bark: reddish brown, thin with many small scales

Cone: green (sometimes purplish), turns reddish brown at maturity, leathery, cylindrical, 1¼–2" (3–5 cm) long, tapers to a round tip, hanging from branch, stalk-less, round cone scales, many winged seeds within, falls apart in autumn

Origin/Age: native; 300–400 years

Habitat: rocky soils on steep slopes, ridges and mountains, acid soils, moist soils

Range: Appalachian Mountains and other mountainous places in the eastern part of the state

Stan's Notes: A tree of the Northeast and the southernmost spruce in the eastern U.S. Often in pure stands. A shallow-rooted tree, falling victim to windstorms. Can thrive in deep shade for many years, but grows nearly twice as fast in full sun. An extremely important commercial tree species. Wood is straight grained, strong, lightweight and used to make musical instruments, lumber and paper. Spruce gum was once used as raw material for chewing gum, and twigs and needles were used with flavoring and sugar to make spruce beer. Starts producing cones and seeds at 20–30 years, but maximum production occurs later. Has very large seed crops every 4–6 years. Hybridizes with Black Spruce (pg. 39) where ranges overlap. Also called Yellow Spruce or Eastern Spruce.

bark

cone

Colorado Spruce
Picea pungens

Family: Pine (Pinaceae)

Height: 40–60' (12–18 m)

Tree: pyramid shape, lower branches are the widest and often touch the ground

Needle: single needle, ½–1" (1–2.5 cm) long, very stiff, very sharp point on the end, square in cross section, bluish green to silvery blue

Bark: grayish brown and flaky, becoming reddish brown and deeply furrowed with age

Cone: straw-colored, 2–4" (5–10 cm) long, in clusters or single, hanging down

Origin/Age: non-native, was introduced to the state from the Rocky Mountains; 150–200 years (can reach 600 years in some western states)

Habitat: variety of soils, does best in clay and moist soils, sun

Range: throughout, planted in cities, parks, along roads and around homes

Stan's Notes: A common Christmas tree and landscaping tree that is widely planted around homes and along city streets. Naturalized now throughout the state. A victim of the Spruce Budworm and needle fungus, so it's not planted as much anymore. Very susceptible to Cytospora canker, which invades stressed trees, causing loss of branches and eventual death. Will grow in a wide variety of soils, but prefers moist and well drained. Slow growing, some living up to 600 years in the West. Needles are very sharp and square in cross section. The species name *pungens* is Latin for "sharp-pointed." Also known as Blue Spruce or Silver Spruce.

cone

bark

Norway Spruce
Picea abies

Family: Pine (Pinaceae)

Height: 50–70' (15–21 m)

Tree: pyramid shape, single trunk, branches drooping or weeping

Needle: single needle, ½–1" (1–2.5 cm) long, with a slight curve, stiff and pointed, square in cross section, aromatic when crushed, deep blue-green

Bark: reddish gray, many round scales

Cone: straw brown, papery, 2–7" (5–18 cm) long, hangs from branch

Origin/Age: non-native, introduced to the U.S. from Europe and Asia; 150–200 years

Habitat: rich moist soils, sun

Range: throughout, planted as windbreaks, in cemeteries, parks and yards

Stan's Notes: Perhaps the most common spruce in Pennsylvania. The fastest growing and one of the tallest spruces in the state, popular for planting as windbreaks. Produces the largest cones of all spruces. Generally a very healthy tree with few diseases. Sometimes has deformed cones, caused by Cooley Spruce Gall Adelgids (aphid-like insects) chewing on its new growth. Introduced from Europe, as the common name implies, it is the dominant tree species in the Black Forest area of Germany. One of the earliest trees used for reforestation in North America. The bark on the twigs is orange, turning reddish brown on the small branches. The trunk oozes a pitch known as burgundy pitch, which has been used in varnishes and medicine. Many horticultural varieties of this tree are available.

bark

cone

Eastern Hemlock
Tsuga canadensis

SINGLE NEEDLE

Family: Pine (Pinaceae)

Height: 40–60' (12–18 m)

Tree: pyramid shape, spreading branches are horizontal with drooping tips, irregular crown

Needle: single needle, ½–1" (1–2.5 cm) long, arranged in 2 rows with a few shorter needles on the upper row, borne on a flexible tan twig, soft, flat and flexible, tapering at the end, dark yellow-green above, lighter-colored with 2 whitish lengthwise parallel lines below

Bark: dark brown to dark gray in color, deeply grooved with broad flat-topped ridges

Cone: green, turning brown at maturity, round to ovate, ½–1" (1–2.5 cm) long, on a short stalk, at the end of twig, hanging down

Origin/Age: native; 150–200 years (some reach 600 years)

Habitat: wet soils, cool moist sites, shade tolerant

Range: throughout

Stan's Notes: One of four species of hemlock in the U.S. and the only one native to Pennsylvania. An extremely long-lived tree, some with trunk diameters measuring 4 feet (1.2 m). A very shade-tolerant tree, often growing in dense shade of taller trees, growing slowly until reaching the canopy. Because the tip of the leader shoot (treetop) droops, it often doesn't grow as straight as the other conifers. Bark is rich in tannic acid (tannin) and was once used to tan hides. Open cones will remain on the tree for up to two years. Has heavy seed crops every 2–3 years. Doesn't reproduce very well, as young trees are fragile and often do not reach maturity. Doesn't transplant well. Also called Canada Hemlock. The state tree of Pennsylvania.

bark

cone

Balsam Fir
Abies balsamea

Family: Pine (Pinaceae)

Height: 50–75' (15–23 m)

Tree: tapering spire, many horizontal branches from the ground up, dark green

Needle: single needle, ½–1" (1–2.5 cm) long, with a spiral arrangement on the twig, soft, flat, blunt-tipped, shiny green above with 2 silvery lengthwise lines or grooves below

Bark: light gray, smooth with many very aromatic raised resin blisters (pitch pockets), breaking with age and leaving brown scales

Cone: bluish, 2–4" (5–10 cm) long, erect in dense clusters near the top of tree

Origin/Age: non-native; 100–150 years

Habitat: moist soils, in shaded forests, along bogs, sun to partial shade

Range: scattered locations in the eastern half of the state, planted throughout Pennsylvania

Stan's Notes: Well known for its fragrant needles, this is a popular Christmas tree because it holds its needles well after cutting. One of 50 fir species worldwide. One of nine fir species in North America and one of only two species east of the Rocky Mountains, with the Fraser Fir (not shown) native to the Appalachian Mountains. Often attacked by the Spruce Budworm, which eats the new needles. The upright cones break apart by autumn, leaving only a thin central stalk. Resin from the trunk was once used for making varnishes and sealing birch bark canoes. The common name "Balsam" comes from the Greek root *balsamon*, which refers to aromatic oily resins found in the tree. Also called Eastern Fir or Canada Balsam.

bark

cone

Table Mountain Pine

Pinus pungens

Family: Pine (Pinaceae)

Height: 20–40' (6–12 m)

Tree: single or multiple crooked trunks, wide irregular crown of horizontal branches

Needle: clustered needles, 2 per cluster, 1¼–2½" (3–6 cm) long, each needle is stout, stiff and slightly twisted, dark green with white stripes

Bark: reddish brown with thin furrows and small scales

Cone: green, turning light brown at maturity, ovate, 2–4" (5–10 cm) long, in clusters of 3–5, often points down toward the trunk, stalkless, cone scales thick and strong, outer edge has a stout curved spine, containing many winged seeds, remaining on the tree many years after opening

Origin/Age: native; 300–400 years

Habitat: dry rocky soils, steep slopes and mountain ridges up to 4,500 feet (1,370 m), sun to partial shade

Range: southeastern quarter of the state

Stan's Notes: This species was first collected and named in 1794 near Table Rock Mountain in North Carolina. Grows only (endemic) in the Appalachian Mountains. Also called Mountain Pine, Prickly Pine or Hickory Pine. Known as Squirrel Pine because squirrels pull unopened cones from the tree to get seeds. The species name *pungens* is Latin for "sharp-pointed" and refers to the sharp spines on cones. Often grows as wide as it is tall. Wood is not straight and has many knots, so it doesn't have much commercial value. Used mainly for firewood and pulpwood. Often the first pine to return after an area is logged or burned. A sticky resin in the cone keeps it closed until exposed to heat from forest fires. Some cones will open if the tree is located on warm southern exposures.

immature
cone

mature cone

bark

Virginia Pine
Pinus virginiana

Family: Pine (Pinaceae)

Height: 30–60' (9–18 m)

Tree: medium tree, semi-straight trunk, long, spreading horizontal branches, irregular round crown

Needle: clustered needles, 2 per cluster, 1½–3" (4–7.5 cm) long; each needle is soft, flexible, slightly twisted and fragrant when crushed, light to dull green

Bark: brown to gray with thin shaggy ridges and flakes

Cone: green, turning reddish brown when mature, egg-shaped, tapers near tip, 1½–2¾" (4–7 cm) long, short stalk, cone scales have a ridge and are tipped with a long prickle, stays on tree for many years

Origin/Age: native; 75–100 years

Habitat: sandy soils, clay, well-drained sites, old fields and abandoned farms, sun

Range: southern half of the state

Stan's Notes: The most common of native pines in the state. It is a medium-sized tree, more short-lived than other pines. Usually grows in pure stands or with Shortleaf Pine (pg. 61), Eastern Redcedar (pg. 69) or other pine species. Often called Poverty Pine or Scrub Pine due to its scrubby, scraggly appearance, a result of the poor soils it inhabits. Frequently used to reforest areas with poor or badly eroded soils. Quickly colonizes recently burned areas and old farm fields. Often planted and sold as a Christmas tree. Noted for its dark green color, pleasant pine scent and retention of needles. Used for pulpwood, firewood and railroad ties. Seldom used in the commercial lumber trade. Favorite of woodpeckers, which excavate nesting cavities in its dead trunks. Its prolific seed source is favored by Pine Siskins and other finches. Deer browse on branches of young trees in winter.

bark

cone

peeling bark

Scotch Pine
Pinus sylvestris

Family: Pine (Pinaceae)

Height: 30–80' (9–24.5 m)

Tree: single trunk that is often crooked, with spreading irregular crown

Needle: clustered needles, 2 per cluster, 1½–3" (4–7.5 cm) long; each needle is stiff, twisted and pointed

Bark: orange-brown and flaky lower, bright orange and papery upper

Cone: ovate, 1–2½" (2.5–6 cm) long, on a short stalk, in clusters of 2–3, frequently pointing backward up the branch

Origin/Age: non-native, introduced to the U.S. from Europe; 100–150 years

Habitat: well-drained sandy soils, sun

Range: throughout, planted along roads, in parks and yards and as shelterbelts

Stan's Notes: One of the more popular Christmas trees grown. Among the first species of trees introduced to North America. The most widely distributed pine in the world, found from Europe to eastern Asia, the Arctic Circle to the Mediterranean Sea, and now North America. In Europe it grows tall and straight, but in North America it seldom has a straight trunk because of the seed source chosen by early settlers; apparently it was easier to collect cones for seeds by climbing trees with crooked trunks. Growing conditions, insect pests and disease also crook trunks. Easily identified by its orange-to-red upper branch bark (see inset) that often peels from the branches in thin papery strips. The main trunk bark has loose scales that fall off to reveal a reddish brown inner bark. Two twisted needles per cluster are characteristic. Also known as Scots Pine.

bark

cone

Austrian Pine
Pinus nigra

Family: Pine (Pinaceae)

Height: 40–60' (12–18 m)

Tree: often irregular-shaped with large, open horizontal branches, broad round crown

Needle: clustered needles, 2 per cluster, 3–6" (7.5–15 cm) long; each needle is twisted, sharply pointed, not breaking cleanly when bent, dark green

Bark: gray-brown with reddish branches, very scaly

Cone: green, turning brown at maturity, woody, ovate, 1–3" (2.5–7.5 cm) long, each cone scale ending in a sharp point

Origin/Age: non-native, introduced to the U.S. from southern Europe; 100 or more years

Habitat: wide variety of soils, sun, shade

Range: throughout, planted in parks, along roads, as windbreaks and wildlife shelterbelts

Stan's Notes: A very important tree, also known as European Black Pine. Originally from Europe, it was introduced to North America in 1759. This was the first species of trees to be planted during the dedication of the Dust Bowl Shelterbelt Project in 1935. Frequently confused with Red Pine (pg. 59) but easily differentiated from it by the way the needles break. Unlike Red Pine needles, Austrian Pine needles don't break cleanly when bent. Widely planted in parks and along roads because of its tolerance to salt spray, air pollution and dry soils. Easily grown from seed, it thrives in many soil types and transplants well.

immature
cone

bark

cone

Red Pine
Pinus resinosa

Family: Pine (Pinaceae)

Height: 40–80' (12–24.5 m)

Tree: single straight trunk, dead lower branches fall off soon after dying, broad round crown

Needle: clustered needles, 2 per cluster, 4–6" (10–15 cm) long; each needle straight, brittle, pointed, breaks when bent, dark green

Bark: reddish brown, becoming redder higher up, many flat scales or plates

Cone: green, turning brown at maturity, 2–3" (5–7.5 cm) long, containing many small brown nutlets

Origin/Age: non-native; 150–200 years

Habitat: dry sandy soils, often in pure stands, sun

Range: throughout, frequently in mass plantings, planted in parks, yards and along streets

Stan's Notes: A very impressive sight when planted in large pure stands. Often planted and sold as Christmas trees. Also called Norway Pine because the early settlers confused the tree with the Norway Spruce (pg. 45) of northern Europe. Often confused with Austrian Pine (pg. 57), which has needles as long but that bend without breaking cleanly. Common name comes from its reddish bark. The scaly bark peels off the mature tree and lies at its base, resembling scattered jigsaw puzzle pieces. Branches occur in whorls around the trunk. Cones remain on tree for several years. Heavy seed crops every 4–7 years. Needs a fire to expose mineral soils for seeds to germinate. Used in reforestation projects.

immature cone

bark

cone

Shortleaf Pine
Pinus echinata

Family: Pine (Pinaceae)

Height: 70–100' (21–30.5 m)

Tree: large tree, single straight trunk, many horizontal branches, older trees often lack branches on lower half, round crown

Needle: clustered needles, 2 or 3 per cluster, 2¾–4½" (7–11 cm) long; each needle is soft, flexible, slender, sharply pointed, yellowish green

Bark: reddish brown with large, irregular flat scales

Cone: yellowish green, turning light brown at maturity, oblong, tapers near tip, 1½–2½" (4–6 cm) long, short stalk, thin cone scale, small prickle at the tip

Origin/Age: native; 100–150 years

Habitat: sandy and gravelly soils on south-facing slopes and ridges, old fields, abandoned farms, sun

Range: south central edge of the state

Stan's Notes: One of the fastest-growing pines. After fires or cutting, it quickly reestablishes with many seedlings and suckering shoots, which is uncommon for pines. Often in pure stands or with other pines. Has a hard, strong, yellow-to-orange wood. Is an important commercial tree, producing lumber, millwork, veneer, pulpwood and flooring. Turpentine is produced from the resin. Has the shortest needles of the major southern yellow pines, hence the common name. Wood is often sold as Southern Yellow Pine. Known by other names such as Carolina Pine, Arkansas Pine, Soft Pine and Bull Pine. Reaches cone-bearing maturity at 20–30 years. Seeds look like small maple seeds and are capable of being carried by the wind as far as a quarter mile from the cone. Widespread in the Southeast, it is native in more than 20 states. Reaches its northern limits in southern Pennsylvania.

bark

cone

Pitch Pine
Pinus rigida

Family: Pine (Pinaceae)

Height: 50–70' (15–21 m)

Tree: single straight trunk, many horizontal branches with large gaps, broad irregular crown

Needle: clustered needles, 3 per cluster, 3–5" (7.5–12.5 cm) long; each is stout, stiff, often twisted, yellow-green

Bark: dark gray to brown, thick and deeply furrowed into broad scales

Cone: yellow-brown, turning light brown when mature, egg-shaped, 1¾–2¾" (4.5–7 cm) long, single on branch, stalkless, each scale is thin, flat and armed with a stiff, curved spine, many winged triangular seeds within, remains on branch for many years

Origin/Age: native; 100–150 years

Habitat: sandy and gravelly soils on steep slopes and ridges with elevations up to 3,500 feet (1,070 m)

Range: throughout, except for the northern quarter

Stan's Notes: The most common native pine in Pennsylvania. The greatest abundance in North America is in Pennsylvania and New Jersey, where it often forms dense pure stands known as Pine Barrens. A pine of nutrient-poor soils and dry locations, often growing in small groups. Slow growing for the first 5–10 years, then grows rapidly. Cones remain unopened on tree for many years. Forest fires cause them to open and release seeds, which helps to colonize newly burned soils. Often used for reforestation where few other trees will grow or where soil has been depleted. Light brown wood is soft, knotty and not very strong but very resistant to decay and mainly used for fuel and charcoal production. Once used as a resin source for making turpentine and tar. Common name refers to the high resin content. Seeds are an important food source for bird species such as Pine Warbler, Pine Grosbeak and chickadee.

bark

immature cone

cone

Eastern White Pine
Pinus strobus

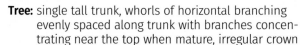

Family: Pine (Pinaceae)

Height: 70–100' (21–30.5 m)

Tree: single tall trunk, whorls of horizontal branching evenly spaced along trunk with branches concentrating near the top when mature, irregular crown

Needle: clustered needles, 5 per cluster, 3–5" (7.5–12.5 cm) long; each needle is soft, flexible and triangular in cross section

Bark: gray to brown and smooth when young, breaking with age into large broad scales that are separated by deep furrows

Cone: green, turning brown when mature, drooping and curved, 4–8" (10–20 cm) long, pointed white tip on each cone scale, resin coated

Origin/Age: native; 200–250 years

Habitat: wide variety of soils, from dry and sandy to moist upland sites, sun

Range: throughout

Stan's Notes: One of the largest conifers in Pennsylvania. A favorite place for Bald Eagles to build their nests. Also known as Northern White Pine, Soft Pine or Weymouth Pine. The most important tree until about 1890 in North America, where its wood was used in buildings in many large eastern U.S. cities. White pine blister rust, a fungus that slowly girdles the trunk, kills many Eastern White Pines. Restoration efforts are underway in many parts of the country to bring this species back.

bark

cone

Tamarack
Larix laricina

Family: Pine (Pinaceae)

Height: 40–70' (12–21 m)

Tree: cone shape, single straight trunk, narrow crown

Needle: clustered needles on any twigs and branches older than 1 year, 12–30 per cluster, ¾–1¼" (2–3 cm) long, single needles on current year's growth; each needle is soft, pointed, triangular in cross section, light green

Bark: gray when young, reddish brown and flaky scales with age

Cone: light brown, round, ½–1" (1–2.5 cm) diameter, on a short curved stalk

Fall Color: bright golden yellow

Origin/Age: non-native; 100–150 years

Habitat: wet soils, swamps, bogs, occasionally in uplands, sun

Range: scattered in central Pennsylvania and the eastern half of the state, planted throughout in parks

Stan's Notes: Like the Bald Cypress (pg. 35), this is a deciduous conifer. A highly unusual species because it sheds its leaves (needles) in autumn. Turns bright golden yellow in the fall before shedding its needles. One of the northernmost trees in North America and also the world. Almost always grows in wetlands but can also be planted as an ornamental in yards. Also known as Eastern Larch or American Larch. Larch Sawfly larvae eat the needles and in some years can defoliate entire stands of Tamarack. The roots of this tree have been used for lashing wooden slats together.

cone

bark

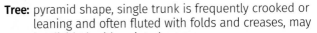

Eastern Redcedar
Juniperus virginiana

Family: Cypress (Cupressaceae)

Height: 25–50' (7.5–15 m)

Tree: pyramid shape, single trunk is frequently crooked or leaning and often fluted with folds and creases, may be divided, with pointed crown

Needle: scaly needles, 1–2" (2.5–5 cm) long, made of scale-like needles, ⅛" (.3 cm) long, that overlap each other, each with a sharply pointed tip, dark green

Bark: reddish brown to gray, thin and fibrous, peeling with age into long narrow shreds; reddish inner bark is smooth

Cone: dark blue with a white powdery film, appearing berry-like, ½" (1 cm) long, containing 1–2 seeds

Fall Color: reddish brown during winter

Origin/Age: native; 300 years

Habitat: dry soils, open hillsides, wet swampy areas, sun

Range: throughout

Stan's Notes: One of the first trees to grow back in prairies or fields after a fire. Slow growing, producing what appear to be blue berries, which are actually cones. Cones are used to flavor gin during the distillation process. Many bird species spread seeds by eating cones, dispersing seeds in their droppings. Redcedar wood is aromatic and lightweight. Often used to make storage chests, lending its scent to linens. The smooth reddish inner bark was called *baton rouge* ("red stick") by early French settlers who found the tree growing in Louisiana. Affected by cedar-apple rust, which causes large jelly-like orange growths. Its sharply pointed leaves can cause slight skin irritation. Also called Eastern Juniper or Red Juniper.

mature
cone

immature
cone

bark

Eastern Whitecedar
Thuja occidentalis

Family: Cypress (Cupressaceae)

Height: 30–50' (9–15 m)

Tree: pyramid shape, single or multiple trunks are often crooked or twisted, blunt or pointed dense crown

Needle: scaly needles, 1–2" (2.5–5 cm) long, made of scale-like needles, ¼" (.6 cm) long, that overlap each other; each scale-like needle is soft, with a rounded tip, flat in cross section, light green

Bark: gray and fibrous with shallow furrows, peeling in long strips

Cone: green, turning bluish purple with a white dust-like coating, then light brown at maturity, ½" (1 cm) long, upright in clusters, containing 2 tiny winged nutlets (seeds)

Origin/Age: native; 150–200 years (some reach 800 years)

Habitat: moist or wet soils, often in pure stands

Range: scattered throughout the state, planted in parks and yards

Stan's Notes: A common tree of bogs and swamps, and a favorite food of deer during winter. Slow growing but with a very long life, with some trees over 700 years old. Also known as Northern Whitecedar, Eastern Thuja or Eastern Arborvitae. The common name "Arborvitae," meaning "tree of life," may have come from French voyagers who used whitecedar to treat scurvy, a disease resulting from a lack of vitamin C. The lightweight wood was once used for canoe frames. One of only two species of *Thuja* in North America, it was introduced into Europe by the mid-1500s. More than 100 different varieties are now known for this tree.

thorn

bark

flower

fruit

European Buckthorn
Rhamnus cathartica

Family: Buckthorn (Rhamnaceae)

Height: 10–20' (3–6 m)

Tree: single or multiple crooked trunks, round crown

Leaf: simple, oval, 1–3" (2.5–7.5 cm) long, oppositely attached, pointed tip, fine-toothed margin, curved and slightly sunken veins, dark green above

Bark: gray, many horizontal white marks (lenticels) and many scales, tiny spine (thorn) in the fork at ends of twigs

Flower: green bell-shaped flower, ¼" (.6 cm) in diameter, in clusters

Fruit: green berry, turning black at maturity, ¼" (.6 cm) in diameter, in clusters, containing 3–4 seeds and remaining on tree throughout winter

Fall Color: green, remaining on tree long after leaves of other trees have dropped

Origin/Age: non-native, introduced from Europe; 25–50 years

Habitat: wide variety of soils, sun to shade

Range: throughout

Stan's Notes: About 100 species of buckthorn trees and shrubs, 12 native to North America. One of two European species that escaped from landscaping. Now naturalized throughout the state. Grows in thick stands, shading out native plants, making it undesirable. Considered a nuisance, many state and city agencies have programs to eliminate it from parks and woodlands. However, the berries are a consistent food supply for birds, which eat them in large quantities and spread the seeds. Berries are cathartic (cause diarrhea) to humans and can cause severe dehydration. Also called Common Buckthorn.

bark

flower

mature fruit

immature fruit

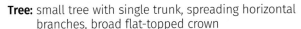

Eastern Flowering Dogwood
Cornus florida

Family: Dogwood (Cornaceae)

Height: 20–30' (6–9 m)

Tree: small tree with single trunk, spreading horizontal branches, broad flat-topped crown

Leaf: simple, oval to round, 2–5" (5–12.5 cm) in length, oppositely attached, smooth toothless margin, prominent curving veins, medium green above, paler below, and covered with fine hairs

Bark: reddish brown, broken into small square plates

Flower: several tiny green flowers, ½" (1 cm) in length, crowded together and surrounded by 4 creamy white bracts, 1–2" (2.5–5 cm) in length, appearing before the leaves in spring

Fruit: green berry-like fruit (drupe), turning red when mature, ¼–½" (.6–1 cm) diameter, several together on a long fruit stalk, each containing 1–2 seeds

Fall Color: red

Origin/Age: native; 30–50 years

Habitat: moist soils, along forest borders, meadows, shade

Range: throughout

Stan's Notes: Early blooming of showy flowers makes this one of the most beautiful flowering trees in Pennsylvania. Fruit is eaten by many bird species. Its hard wood is used to make mallet handles and heads. Red dye was once extracted from the roots. Often planted as an ornamental for its flowers and bright red fall foliage, and as food for wildlife. Because daggers and meat skewers were made from its wood, common name "Daggerwood" may have been slightly changed to "Dogwood."

bark

flower

fruit

Eastern Wahoo

Euonymus atropurpureus

Family: Staff-tree (Celastraceae)

Height: 20–25' (6–7.5 m)

Tree: single or multiple trunks, irregular crown

Leaf: simple, oval, 2–5" (5–12.5 cm) long, oppositely attached, pointed tip, fine-toothed margin, dull green with hairy underside

Bark: greenish gray with reddish-brown streaks, smooth

Flower: 4-parted purple flower

Fruit: 4-lobed capsule, turning pink to red at maturity, ½" (1 cm) long, containing 4 seeds

Fall Color: red

Origin/Age: native; 25–50 years

Habitat: moist soils, usually is found along streams, rivers and floodplains, partial shade

Range: throughout

Stan's Notes: Also called Spindle Tree or Burning-bush Euonymus. This is the only *Euonymus* tree species that is native to the state. The pinkish capsules each contain four seeds, which have a fleshy covering. Capsules stay on the tree into winter, with the fleshy fruit and seeds providing a good food source for birds. Seeds are spread by birds. Twigs have distinctive corky ridges or wings. Over 150 species in the world, nearly all in Asia, with one tree and four shrub species in North America. Several of the shrubby *Euonymus* species that were introduced are also called Burning-bush.

fruit

bark

flower

leafstalk

Nannyberry
Viburnum lentago

Family: Moschatel (Adoxaceae)

Height: 10–20' (3–6 m)

Tree: appearing like a large shrub with multi-stemmed trunk, drooping branches, and dense round crown

Leaf: simple, oval, 3–5" (7.5–12.5 cm) long, oppositely attached, pointed tip, fine-toothed margin, shiny green, leafstalk (petiole) is flattened to U-shaped with many swellings (glands)

Bark: gray and smooth, many horizontal lines (lenticels), sometimes with small scales

Flower: white flower, ¼" (.6 cm) wide, in flat clusters, 3–5" (7.5–12.5 cm) wide

Fruit: green berry-like fruit, turning dark purple when mature, appearing like a raisin, sweet and edible, round, ¼–½" (.6–1 cm) diameter, in clusters, has 1 seed within

Fall Color: red to reddish purple

Origin/Age: native; 10–20 years

Habitat: wide variety of soils from wet to dry, along forest edges, on stream banks and hillsides

Range: scattered throughout the state

Stan's Notes: Over 100 species in the world with over 20 native to the U.S. Also known as Blackhaw, Sweet Viburnum, or Wild Raisin, the last name referring to the taste and texture of the fruit when ripe in late summer. Although the fruit is sweet and edible, the large seed within makes it hard to eat. A favorite food of wildlife when ripe. Grooved or winged leafstalk (see inset) helps to identify the tree. Often along edges of forest, swamps, marshes and streams.

bark

flower

fruit

Buttonbush
Cephalanthus occidentalis

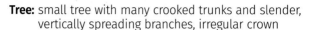

Family: Madder (Rubiaceae)

Height: 10–20' (3–6 m)

Tree: small tree with many crooked trunks and slender, vertically spreading branches, irregular crown

Leaf: simple, oval, 2–6" (5–15 cm) long, often opposite but can be in whorls of 3, pointed at tip, round at base, wavy toothless margin, semi-evergreen, shiny green above, paler below, sometimes covered with fine hairs

Bark: gray to brown, deeply furrowed with scaly ridges

Flower: many small white flowers, each ½–¾" (1–2 cm) long, in balls, 1–1½" (2.5–4 cm) wide, on stalks

Fruit: round green aggregate of many small nutlets, turning brown at maturity, ¾–1" (2–2.5 cm) wide, on a long fruit stalk

Fall Color: brown, but can be green

Origin/Age: native; 10–20 years

Habitat: moist soils, along streams and lakes, floodplains, wet meadows, shade

Range: throughout

Stan's Notes: A lowland species found growing in wet areas. Often in association with other lowland tree species such as cottonwood and willow. Sometimes in dense pure stands, usually on lakeshores. The dense stands often make good cover for nesting birds. A fast-growing, short-lived semi-evergreen with attractive flower balls. Also known as Honey-balls or Globe-flowers. Produces large quantities of nutlets, which mature in fall and are eaten by wildlife such as ducks, deer and turkeys. Foliage and bark are poisonous and have been used in folk medicine, but the effectiveness is highly doubtful.

bark

flower

fruit

Northern Catalpa
Catalpa speciosa

Family: Trumpet-Creeper (Bignoniaceae)

Height: 50–75' (15–23 m)

Tree: single trunk, large round crown

Leaf: simple, heart-shaped, 6–12" (15–30 cm) in length, oppositely attached or whorls of 3 leaves, margin lacking teeth, dull green

Bark: light brown with deep furrows, flat-topped ridges

Flower: large, showy orchid-like flower is cream to white with yellow and purple spots and stripes, 2–3" (5–7.5 cm) long, in clusters, 5–8" (13–20 cm) wide, fragrant scent

Fruit: long bean-like green capsule, turning to brown at maturity, 8–18" (20–45 cm) long, splitting open into 2 parts, containing winged seeds

Fall Color: yellow-green, turning black

Origin/Age: non-native, was introduced to the state from the central Mississippi Valley; 40–50 years

Habitat: rich moist soils

Range: isolated in cities, parks and backyards throughout the state, old homesites

Stan's Notes: Catalpa tree leaves are among the largest leaves in the state. There are about a dozen catalpa species, two native to the U.S. This is a non-native tree that has been successfully planted along streets and boulevards in just about every city in the state. "Catalpa" is an American Indian name for this tree, but it is also called Catawba, Cigar Tree or Indian Bean, which all refer to the large seedpods (fruit). Its large showy flowers bloom in spring and attract many insects. Twigs have a soft white pith.

fruit

bark

Siberian Elm
Ulmus pumila

Family: Elm (Ulmaceae)

Height: 30–50' (9–15 m)

Tree: single trunk, open irregular crown

Leaf: simple, narrow, ¾–2" (2–5 cm) in length, alternately attached, with pointed tip, asymmetrical leaf base, double-toothed margin, dark green

Bark: gray with rough scales

Fruit: flat green disk (samara), lacking hair when young, turning papery brown when mature, ½" (1 cm) diameter, with a closed notch opposite fruit stalk

Fall Color: yellow

Origin/Age: non-native, introduced from Asia; 50–75 years

Habitat: wide variety of soils, sun

Range: throughout, in parks and yards, as hedges, around old homesites

Stan's Notes: A small non-native species that was introduced from Asia, with some of the smallest leaves of any of the elm trees. Fast growing but does not attain the height or reach the age of American Elm (pg. 87). Species name *pumila* means "small" and refers to the small stature. Also called Chinese Elm. Chinese Elm (*U. parvifolia*), however, is a different, cultivated species. Thrives in a wide variety of soils and tolerates harsh conditions. Somewhat resistant to Dutch elm disease, often taking much longer to die from the disease than the American Elm, which usually dies quickly.

bark

flower

fruit

American Elm
Ulmus americana

Family: Elm (Ulmaceae)

Height: 70–100' (21–30.5 m)

Tree: one of the tallest of trees, single trunk, prominent root flares, upper limbs fan out gracefully, forming an upright vase shape, branch tips often drooping

Leaf: simple, oval, 3–6" (7.5–15 cm) in length, alternately attached, with pointed tip, asymmetrical leaf base, double-toothed margin, slightly rough to touch, only 2–3 forked veins per leaf

Bark: dark gray, deeply furrowed with flat ridges, corky, sometimes scaly

Flower: tiny reddish-brown flower, ¼" (.6 cm) diameter, in clusters, 1" (2.5 cm) wide

Fruit: flat, fuzzy green disk (samara), turning tan when mature, round to oval, ½" (1 cm) diameter, with a notch opposite the fruit stalk

Fall Color: yellow

Origin/Age: native; 150–200 years

Habitat: moist soils, full sun

Range: throughout, planted along city streets, parkways

Stan's Notes: Also called White Elm, this once dominant tree lined just about every city street in eastern North America. Nearly eliminated due to Dutch elm disease, which is caused by a fungus that attacks the tree's vascular system. The fungus was introduced to the U.S. in the 1920s by infected elm logs from Europe. Its arching branches form a canopy, providing shade. The distinct vase shape of the mature tree makes it easy to recognize from a distance. Several Dutch elm disease-resistant trees are now sold at nurseries.

bark

fruit

Slippery Elm
Ulmus rubra

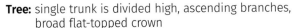

Family: Elm (Ulmaceae)

Height: 50–70' (15–21 m)

Tree: single trunk is divided high, ascending branches, broad flat-topped crown

Leaf: simple, oval, 4–7" (10–18 cm) in length, alternately attached, widest above the middle, asymmetrical leaf base, approximately 26–30 veins with some forked near margin, rough and dark green above, paler below, short leafstalk

Bark: brown to reddish brown, shallow furrows, vertical irregular flat scales, inner bark reddish

Fruit: green disk (samara), turning brown when mature, nearly round, ½–¾" (1–2 cm) in diameter, with a slightly notched tip opposite the fruit stalk, few reddish-brown hairs

Fall Color: yellow

Origin/Age: native; 100–125 years

Habitat: rich moist soils, along streams, slopes, sun

Range: throughout

Stan's Notes: Due to its habit of growing near water, sometimes is called Water Elm. More often called Red Elm because of its reddish inner bark. The inner bark is fragrant and mucilaginous, hence its common name, "Slippery." Scientific name was formerly *U. fulva*. The inner bark was once chewed to quench thirst and used to cure sore throats. Leaves are smaller than those of most other elm species and feel very rough. Look for some forked veins near the margin. Leaf buds are dark rusty brown and covered with hairs, making this tree easy to identify even without its leaves.

bark

flower

fruit

Quaking Aspen
Populus tremuloides

SIMPLE ALTERNATE

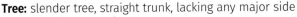

Family: Willow (Salicaceae)

Height: 40–70' (12–21 m)

Tree: slender tree, straight trunk, lacking any major side branches, round crown

Leaf: simple, nearly round, 1–3" (2.5–7.5 cm) in length, alternately attached, with short sharp point, fine-toothed margin, shiny green above and dull green below, leafstalk (petiole) flattened

Bark: dark gray to brown in color and deeply furrowed lower, greenish white to cream and smooth upper

Flower: catkin, 1–2" (2.5–5 cm) in length, male and female bloom (flower) on separate trees (dioecious) in spring before the leaves bud

Fruit: catkin-like fruit, 4" (10 cm) long, is composed of many tiny green capsules, ⅛" (.3 cm) long, that open and release seeds, seeds are attached to white cottony material and float on the wind

Fall Color: golden yellow

Origin/Age: native; 60–80 years

Habitat: wet or dry, sandy or rocky soils, sun

Range: throughout

Stan's Notes: The most widely distributed tree in North America. Common name refers to the leaves, which catch very gentle breezes and shake or quake in the wind. Also known as Trembling Aspen or Popple. Grows in large, often pure stands. Returns from its own roots if cut or toppled. Most reproduce by suckering off their roots, which creates clone trees. One stand in Utah—106 acres (42 ha) in size, with about 47,000 trunks—is the largest single living organism by mass in the world. Has survived lab temperatures of –314°F (–192°C).

bark

fruit

Bigtooth Aspen
Populus grandidentata

SIMPLE ALTERNATE

Family: Willow (Salicaceae)

Height: 50–70' (15–21 m)

Tree: single trunk with few lower branches and a round irregular crown

Leaf: simple, oval to triangular, 3–6" (7.5–15 cm) long, alternately attached, blunt tip, with up to 30–34 large saw-like blunt teeth, waxy above, has a long flattened leafstalk

Bark: gray in color with deep furrows and thick ridges lower, pale green to white and smooth upper

Flower: catkin, 4–5" (10–12.5 cm) long

Fruit: catkin-like fruit, 4" (10 cm) long, is composed of many tiny narrow capsules, ⅛" (.3 cm) long, that split open into 2 parts and release cottony seeds

Fall Color: yellow

Origin/Age: native; 50–75 years

Habitat: moist soils, sun

Range: throughout

Stan's Notes: One of over 40 species of poplar found throughout the northern hemisphere. A fast-growing, short-lived tree. Woolly hairs cover leaves in spring, but by summer they will have turned thick and waxy. Also called Largetooth Aspen, as each leaf often has up to 30–34 large teeth. Leafstalk (petiole) is flattened and often as long as the leaf. Its large leaves catch the slightest breezes, causing rustling in the same way as the Quaking Aspen (pg. 91). Male and female flowers grow on separate trees (dioecious). Produces heavy seed crops every 3–5 years.

bark

fruit

Lombardy Poplar
Populus nigra

Family: Willow (Salicaceae)

Height: 60–80' (18–24.5 m)

Tree: single straight trunk, ascending branches, narrow columnar crown

Leaf: simple, triangular, 2–4" (5–10 cm) long, alternately attached, often wider than long, pointed tip, fine-toothed margin, dark green above, paler below

Bark: gray with deep furrows lower and smooth upper

Fruit: capsule, ¼" (.6 cm) long, containing many seeds

Fall Color: yellow to brown

Origin/Age: non-native, introduced from Italy; 30–50 years

Habitat: wide variety of soils, sun

Range: throughout, planted as windbreaks and as an ornamental in yards and parks

Stan's Notes: A uniquely tall, thin tree, usually seen in cities, parks or near older homesteads. Unable to reproduce because all trees planted are males. Often planted in a single row, it was once planted much more than it is planted now. Removed from many city parks because of its non-native status. Fast growing, but short-lived due to its susceptibility to various canker diseases. The species name *nigra* refers to its dark patchy bark. The common name "Lombardy" is also the name of a region in Italy. Known for lining the streets of many Italian cities. Also known as Black Poplar.

underside

bark

flower

fruit

Balsam Poplar
Populus balsamifera

Family: Willow (Salicaceae)

Height: 50–70' (15–21 m)

Tree: single trunk with ascending branches and narrow open crown

Leaf: simple, triangular, 3–6" (7.5–15 cm) long, alternately attached, fine-toothed margin, shiny green above, silvery green below with resinous, fragrant rust-colored blotches, round leafstalk

Bark: greenish-brown color when young, aging to gray, smooth with many cracks and fissures

Flower: catkin, 3–4" (7.5–10 cm) long

Fruit: catkin-like fruit, 3–4" (7.5–10 cm) long, composed of many tiny capsules, ⅛" (.3 cm) long, that split open into 2 parts and release seeds, which are attached to cottony hair and float on the wind

Fall Color: yellow

Origin/Age: native; 50–75 years

Habitat: moist soils, river valleys, shade intolerant

Range: scattered in pockets throughout the state

Stan's Notes: It's easy to smell this tree—simply walk near it. Small branches pruned during springtime can be brought inside for a wonderful spicy fragrance. In the spring, its leaf buds are covered with a sticky, fragrant resin. Later, the rust-colored resin covers the undersides of leaves. Fast growing, shade intolerant, often growing in pure stands or mixed with aspens. Its species name is Latin and refers to its scent. Also known as Balm-of-Gilead, which refers to the alleged medicinal properties of the resin.

bark

flower

seeds

Eastern Cottonwood

Populus deltoides

Family: Willow (Salicaceae)

Height: 70–100' (21–30.5 m)

Tree: large tree with single or multiple trunks, few lower branches and huge, broad irregular crown

Leaf: simple, triangular, 3–6" (7.5–15 cm) in length, alternately attached, coarse-toothed margin, thick and waxy, shiny green above and below, leafstalk long and flattened

Bark: gray with deep flat furrows

Flower: catkin, 2–3" (5–7.5 cm) long

Fruit: catkin-like fruit, 4" (10 cm) long, is composed of many tiny ovate capsules, ¼" (.6 cm) long, that split open into 4 parts and release seeds, seeds (see inset) are attached to cotton-like filaments and float on wind

Fall Color: yellow

Origin/Age: native; 50–200 years

Habitat: wet soils, along streams, rivers and lakes, sun

Range: scattered throughout the western half of the state

Stan's Notes: A huge tree of riverbanks, floodplains and other wet areas. Some trees can obtain heights of 150 feet (46 m), with trunk diameters of 7–8 feet (2.1–2.4 m). Fast growing, up to 5 feet (1.5 m) in height and over 1 inch (2.5 cm) in diameter per year. Like many others in its genus, the leafstalks are flat. The species name *deltoides* is Latin, describing the delta-shaped leaf. Known for the massive release of seed-bearing "cotton," hence its common name.

fruit

bark

Gray Birch

Betula populifolia

Family: Birch (Betulaceae)

Height: 10–30' (3-9 m)

Tree: small tree, single or multiple crooked or leaning trunks, many short branches, conical crown

Leaf: simple, triangular, 2–3½" (5–9 cm) length, alternately attached, a long, tapering, pointed tip, sharp double-toothed margin, papery in texture, shiny dark green, slender leafstalk, ½–1" (1–2.5 cm) long, allowing the leaf to hang vertically

Bark: gray to chalky white with dark horizontal marks, thin and non-peeling

Flower: catkin, ¾–1¼" (2–3 cm) long

Fruit: pair of winged nutlets, ⅛" (.3 cm) wide, fall from cone-like seed catkin in autumn and winter

Fall Color: yellow

Origin/Age: native; 50–75 years

Habitat: nutrient-poor soils, moist soils, dry open uplands

Range: eastern quarter of the state and scattered locations elsewhere in Pennsylvania

Stan's Notes: Fast-growing tree that reaches reproductive maturity at about 10 years. Considered a pioneer tree, often growing in pure stands in abandoned fields and recently burned forests. The entire tree often leans or bends over with the upper branches reaching the ground, especially during winter when under heavy snow. Its long stalked leaves shake and tremble in the slightest wind. New leaves, flowers and seeds are eaten by grouse and other birds and wildlife. Its reddish-brown wood is light, soft and has little commercial value, but has been used for turning spools and firewood. Also called Wire Birch or White Birch.

bark

fruit

Paper Birch
Betula papyrifera

Family: Birch (Betulaceae)

Height: 40–60' (12–18 m)

Tree: single or multiple crooked trunks with drooping branches and open narrow crown

Leaf: simple, oval to triangular, 2–4" (5–10 cm) in length, alternately attached, pointed tip, double-toothed margin, 18 or fewer veins, each ending in a large tooth, dull green above, paler below

Bark: white and smooth with obvious dark horizontal lines (lenticels), often shedding in curled sheets, inner bark reddish

Flower: catkin, 1–2" (2.5–5 cm) long

Fruit: many winged nutlets, each ⅛" (.3 cm) wide, in a cone-like seed catkin, 1–2" (2.5–5 cm) long

Fall Color: yellow

Origin/Age: native; 80–100 years

Habitat: moist or sandy soils, partial to full shade

Range: scattered throughout the state, planted throughout as an ornamental tree

Stan's Notes: Also called White Birch or Canoe Birch, both names referring to the white bark that American Indians have long used to construct canoes, baskets and water containers. Dried bark is often used to start campfires. Understory tree that likes moist soil and high humidity. Often doesn't grow well when planted by itself in a sunny suburban lawn. When stressed, trees are attacked by destructive Bronze Birch Borer beetle larvae. Both the male and female flowers are on the same tree (monoecious). Winged nutlets are released from seed catkins in early winter. Found across the northern portion of North America.

immature
fruit

bark

fruit

Yellow Birch
Betula alleghaniensis

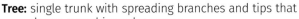

Family: Birch (Betulaceae)

Height: 50–70' (15–21 m)

Tree: single trunk with spreading branches and tips that droop, round irregular crown

Leaf: simple, oval to lance-shaped, 3–5" (7.5–12.5 cm) in length, alternately attached, pointed tip, double-toothed margin, dull green

Bark: bronze to yellow in color, thin horizontal marks (lenticels), covered with thin papery scales often curling up in rolls

Flower: catkin, 1–2" (2.5–5 cm) long

Fruit: many winged nutlets, ⅛" (.3 cm) wide, contained in a cone-like seed catkin, 1" (2.5 cm) long, that grows upright on branch

Fall Color: yellow

Origin/Age: native; 100–125 years

Habitat: rich moist soils, often in wet places, partial shade

Range: throughout

Stan's Notes: One of the tallest of the birches, often thought to be underutilized in landscaping. Also called Swamp Birch due to its nature of growing in wet areas. Common name Yellow Birch refers to the color of the tree's bark. The bark is very characteristic, making this species one of the easier birches to identify. Twigs are aromatic of wintergreen (methyl salicylate) when crushed. A pleasant-tasting tea can be made from the tender twigs. Wood is used to make furniture and veneers. Will produce heavy seed crops every couple of years. Seedlings are heavily browsed by deer.

fruit

bark

flower

Crab Apple
Malus spp.

Family: Rose (Rosaceae)

Height: 10–20' (3–6 m)

Tree: single crooked trunk, broad open crown

Leaf: simple, oval, 2–3" (5–7.5 cm) in length, alternately attached, sometimes with shallow lobes, double-toothed margin, dark green above, lighter-colored and usually smooth or hairless below

Bark: gray, many scales, with 1–2" (2.5–5 cm) long stout thorns often on twigs

Flower: 5-petaled white-to-pink or red flower that is often very showy, 1–2" (2.5–5 cm) wide

Fruit: apple (pome), ranging in color from green and yellow to red, edible, 1–3" (2.5–7.5 cm) diameter, single or in small clusters, hanging from a long fruit stalk well into winter

Fall Color: yellow to red

Origin/Age: native and non-native; 25–50 years

Habitat: wide variety of soils, sun

Range: throughout, often around cities or old homesites

Stan's Notes: Many species of cultivated Crab Apple can be found throughout the state. Others have escaped cultivation and now grow in the wild. Introduced to the U.S. in colonial times. Has since bred with native species, producing hybrids that are hard to identify. Now found throughout the country. Apples are closely related to those sold in grocery stores and have been used in jams and jellies. Cider is often made from the more tart apples. Fruit is an important food source for wildlife. Twigs often have long stout thorns, which are actually modified branches known as spur branches.

bark

flower

fruit

Wild Apple

Malus spp.

Family: Rose (Rosaceae)

Height: 10–15' (3–4.5 m)

Tree: single crooked trunk, many spreading branches, creating a broad round crown

Leaf: simple, oval, 2–4" (5–10 cm) in length, blunt-tipped, fine-toothed margin, dark green in color, densely hairy below

Bark: brown, scaly with peeling edges

Flower: 5-petaled showy white (sometimes streaked with pink) flower, 1–2" (2.5–5 cm) wide

Fruit: apple (pome), edible with typical shape and size, 2–4" (5–10 cm) diameter

Fall Color: brown

Origin/Age: non-native; 25–50 years

Habitat: dry soils, along fencerows and roadsides, sun

Range: throughout

Stan's Notes: Common apples sold in grocery stores are from trees descended from the Wild Apple. Introduced to the U.S. in colonial times, like many varieties of Crab Apple (pg. 107). The Wild Apple, often associated with former homesteads, is found along roads or fencerows where seedlings were planted or where apples were discarded and seeds have taken root. Some escaped cultivation, and now many varieties are naturalized throughout the country. Wild apples are edible and some are very delicious. The fruit has been used in jellies and desserts, such as pies. A wide variety of Wild Apple species are now naturalized in Pennsylvania.

underside

bark

flower

fruit

Weeping Willow
Salix babylonica

Family: Willow (Salicaceae)

Height: 40–80' (12–24.5 m)

Tree: single trunk, often crooked, with many drooping branches often reaching the ground, very broad round crown

Leaf: simple, narrow lance-shaped, 2–4" (5–10 cm) long, alternately attached, pointed tip and fine-toothed margin, bright green above, whitish below

Bark: light brown with many medium furrows and flat-topped corky ridges

Flower: catkin, 1" (2.5 cm) long, standing erect on a short leafy shoot

Fruit: catkin-like fruit, 1" (2.5 cm) long, is composed of many small capsules that open and release seeds, which are attached to white cottony material

Fall Color: yellow

Origin/Age: non-native, introduced from Asia; 75–100 years

Habitat: wet or moist soils, sun

Range: throughout, planted as an ornamental in parks and yards

Stan's Notes: Also known as Golden Willow because of its yellow twigs and autumn color. Twigs are very flexible and hang nearly to the ground. Many branches break off in heavy winds. Often seen growing along shores of ponds and lakes. Not as commonly planted anymore. An extract from willow bark (salicin) is related to aspirin and was once used in a similar manner. Several tree species have been sold under the name of Weeping Willow, each only slightly different in leaf size or stem color (*S. alba, S. pendulina*). Also called Babylon Weeping Willow.

fruit

bark

flower

Pussy Willow

Salix discolor

Family: Willow (Salicaceae)

Height: 10–20' (3–6 m)

Tree: small tree, multiple trunks, irregular crown

Leaf: simple, lance-shaped, 2–5" (5–12.5 cm) in length, alternately attached, with shallow irregular teeth, smooth to touch, shiny green above, paler below, lacking hairs, young leaves often reddish and very hairy, 2 small appendages (stipules) on leafstalk

Bark: gray to brown, often red tinges, shallow furrows

Flower: catkin, ½–1" (1–2.5 cm) long, covered with silky hairs when immature, lacking a stalk

Fruit: catkin-like fruit, ½–1" (1–2.5 cm) long, composed of many capsules, ¼" (.6 cm) long, seeds within

Fall Color: yellow

Origin/Age: native; 20–50 years

Habitat: wet soils, along shores, swamps, wetlands, sun

Range: throughout

Stan's Notes: This is one of the best-known native willows in the state. The immature catkins, which are covered with silky hairs known as pussy fur, are often collected and used in floral arrangements in early spring. Twigs are reddish purple with orange dots (lenticels). Species name *discolor* refers to the pale underside of leaves. Look for tiny leaf-like stipules on the leafstalk and contrasting green upper and whitish lower leaf surfaces to help identify.

bark

fruit

Black Willow
Salix nigra

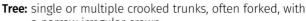

Family: Willow (Salicaceae)

Height: 40–60' (12–18 m)

Tree: single or multiple crooked trunks, often forked, with a narrow irregular crown

Leaf: simple, narrow lance-shaped, 3–6" (7.5–15 cm) in length, alternately attached, fine-toothed margin, shiny green above and below, short leafstalk often has tiny leaf-like appendages (stipules), making the leaf appear to be clasping the twig

Bark: dark brown color and deeply furrowed into large patchy scales with flat-topped ridges

Flower: catkin, 2–3" (5–7.5 cm) long, hanging down

Fruit: catkin-like fruit, 2–3" (5–7.5 cm) long, composed of many capsules, ¼" (.6 cm) long, seeds within

Fall Color: light yellow

Origin/Age: native; 50–75 years

Habitat: wet soils, stream banks, wetlands and other wet places, shade intolerant

Range: throughout

Stan's Notes: The largest native willow in Pennsylvania. Common name comes from the dark bark. A fast-growing but short-lived tree that does not tolerate shade. Commonly found along streams, rivers and other wet places. Also called Swamp Willow. Frequently hybridizes with the Crack Willow (*S. fragilis*), a non-native European species. Twigs are light yellow to reddish and downy when young, becoming gray and hairless. Branches are spreading and easily broken by high winds. Look for its large leaves and leaf-like stipules to help identify.

thorn

bark

flower

fruit

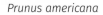

American Plum
Prunus americana

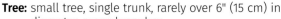

Family: Rose (Rosaceae)

Height: 10–15' (3–4.5 m)

Tree: small tree, single trunk, rarely over 6" (15 cm) in diameter, many branches

Leaf: simple, oblong to oval, 2–5" (5–12.5 cm) in length, alternately attached, pointed tip, double-toothed margin, network of veins, dark green

Bark: reddish brown or gray to white in color with large light-colored horizontal marks (lenticels), smooth, breaking into large scales or plates, large sharp thorns, 1–3" (2.5–7.5 cm) long

Flower: showy white flower, 1" (2.5 cm) wide, in clusters, 3–5" (7.5–12.5 cm) wide, very fragrant

Fruit: large fleshy red plum (drupe), edible, 1" (2.5 cm) wide, containing 1 large seed

Fall Color: golden yellow

Origin/Age: native; 25–30 years

Habitat: moist soils, in open fields, along woodland edges, shade intolerant

Range: throughout

Stan's Notes: A fast-growing ornamental tree also known as Wild Plum. Shade intolerant, preferring open sunny sites. Flowers are large and attractive. Resulting plums are highly prized by wildlife and people. Plums make good jellies and jams; can be eaten fresh. Twigs often smell of bitter almond when they are crushed. The large sharp thorns (see inset) are actually modified branches. In pioneer times, its thorns were used for mending clothes and other tasks.

bark

flower

fruit

Pin Cherry
Prunus pensylvanica

SIMPLE ALTERNATE

Family: Rose (Rosaceae)

Height: 10–30' (3–9 m)

Tree: single straight or crooked trunk, narrow crown

Leaf: simple, lance-shaped, 2–5" (5–12.5 cm) in length, alternately attached, frequently curved backward, tapering to a point, fine-toothed and often wavy margin, shiny green above and below, 2 small swellings (glands) on the leafstalk near leaf base, leaves often clustered at ends of branches

Bark: gray to nearly black in color, smooth and shiny when young, brown-to-orange marks (lenticels) with age, sometimes peeling into papery strips, pleasant scent when scraped

Flower: 5-petaled white flower, ½" (1 cm) wide, on a long stalk, in clusters, 1–2" (2.5–5 cm) long

Fruit: green cherry (drupe), turning bright red when mature, edible, ¼" (.6 cm) diameter, on a very long red fruit stalk

Fall Color: purplish red

Origin/Age: native; 20–40 years

Habitat: dry soils, open hillsides, fields, sun

Range: throughout

Stan's Notes: Also called Wild Red Cherry, Bird Cherry or Fire Cherry. Common name "Fire" refers to its habit of growing quickly after fires. An important pioneer species, invading openings created by logging, fire or abandoned fields. A great food source for wildlife. Its cherries are especially good in jellies and were once used in cough medicines. Long stalks of the flowers and fruit help differentiate it from the Choke Cherry (pg. 121) and Black Cherry (pg. 123).

fruit

bark

flower

fruit

Choke Cherry
Prunus virginiana

Family: Rose (Rosaceae)

Height: 15–35' (4.5–11 m)

Tree: several crooked trunks, irregular crown is usually open with many branches missing

Leaf: simple, oval, 2–5" (5–12.5 cm) length, alternately attached, often is widest above the middle, short sharp tip, fine-toothed margin, shiny green above, lighter below, 2 small swellings (glands) near leafstalk (petiole) near leaf base

Bark: dark brown to gray, smooth texture, occasionally with scales, rank odor when scraped or crushed

Flower: 5-petaled white flower, ½" (1 cm) wide, in spike clusters, 2–3" (5–7.5 cm) long, unpleasant odor

Fruit: yellow-to-red cherry, turning nearly black when mature, ¼" (.6 cm) diameter, 6–20 per hanging cluster, ripening in late summer, extremely bitter

Fall Color: yellow, reddish

Origin/Age: native; 25–50 years

Habitat: wide variety of soils, often along fencerows and streams, woodland edges

Range: throughout

Stan's Notes: Appears like a large shrub. Covered with flowers in spring and bitter red-to-black cherries in late summer. Entire plant, except for soft parts of the fruit, contains cyanide, which smells and tastes like bitter almond. Flowers have five petals, like other Rose family members. Birds and other animals eat the fruit. Spread by birds, which void seeds while perching on fences, and by suckers growing from roots, which create large stands. Lacks brown hairs on midrib, as seen on Black Cherry (pg. 123) leaf undersides.

midrib hairs

immature fruit

bark

flower

fruit

Black Cherry
Prunus serotina

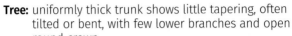

Family: Rose (Rosaceae)

Height: 50–75' (15–23 m)

Tree: uniformly thick trunk shows little tapering, often tilted or bent, with few lower branches and open round crown

Leaf: simple, lance-shaped, 2–6" (5–15 cm) long, alternately attached, with a unique inward-curved tip resembling a bird's beak, fine-toothed margin, row of fine brown hairs along midrib underneath (see inset), shiny dark green above, paler below

Bark: dark reddish brown to black in color with large, conspicuous curving scales (like potato chips), green inner bark tastes bitter but smells pleasant

Flower: white flower, ½" (1 cm) wide, 6–12 per elongated cluster, 4–6" (10–15 cm) long

Fruit: green cherry (drupe), red to dark blue or black at maturity, edible, ¼–½" (.6–1 cm) wide, in clusters

Fall Color: yellow

Origin/Age: native; 125–150 years

Habitat: wide variety soils, mixed with deciduous species

Range: throughout

Stan's Notes: Largest of the cherry trees. Widely sought for its wonderfully rich brown wood. Several species of *Prunus* in Pennsylvania, this one producing a tart but edible fruit. An important food crop for birds and wildlife. Bark and roots contain hydrocyanic acid, which has been used in cough medicines and for flavoring. Black knot (a fungus) is the most common disease of Black Cherry, resulting in large black growths along twigs and small branches, leading to die-back of affected branches.

bark

fruit

Shingle Oak
Quercus imbricaria

Family: Beech (Fagaceae)

Height: 50–60' (15–18 m)

Tree: medium tree, single straight trunk, wide-spreading branches, wide round crown

Leaf: simple, oval to oblong (sometimes lance-shaped), 3–6" (7.5–15 cm) in length, alternately attached, pointed tip, edges wavy or slightly turned under, lacking teeth, shiny dark green above, light green with soft hairs below, midrib often yellowish

Bark: brown to gray, turning darker and more furrowed with age, thinner than the bark of most other oaks

Fruit: green acorn, turning brown when mature, edible, round, ½–¾" (1–2 cm) long, single or in pairs on a short stalk, cap covering the upper third to half of nut, maturing in 2 seasons

Fall Color: yellow to reddish brown

Origin/Age: native; 100–150 years

Habitat: moist soils, along streams and rivers, higher dry soils, sun

Range: throughout

Stan's Notes: Often planted as an ornamental tree and for shade due to its large attractive foliage and tolerance of soil conditions. One of the most abundant oaks in the Ohio River Valley. Moderate to fast growing, medium life span. Usually in moist sites with non-mature forests. Flowers in spring. Acorns mature in fall of second year and are an important food for waterfowl, such as Mallards and Wood Ducks. Wood is easy to split and was once used for roofing shingles, hence its common name. Wood is now sold as Red Oak. The only oak in the state with smooth-edged leaves (lacking lobes or coarse teeth). Holds its leaves into winter, making it a good windbreak.

bark

fruit

Chinquapin Oak
Quercus muehlenbergii

Family: Beech (Fagaceae)

Height: 50–70' (15–21 m)

Tree: medium tree, single straight trunk, narrow round crown composed mostly of many short thin twigs

Leaf: simple, 4–6" (10–15 cm) long, alternately attached, with pointed tip, many straight parallel veins, each ending in a curved coarse tooth, shiny dark green above, pale green and slightly hairy below

Bark: light gray to brown, many long narrow scales

Fruit: green acorn, turning brown when mature, ½–1" (1–2.5 cm) long, on a short stalk, cap covering the upper third of nut

Fall Color: red or brown

Origin/Age: native; 100–150 years

Habitat: sandy soils, dry rocky outcroppings, sun

Range: throughout the southern half of the state

Stan's Notes: Very common tree in the state, occuring in limestone rock outcroppings, ridges and other dry rocky sites. Common name comes from its leaves, which resemble those of trees in a group known as chinquapin. Also called Yellow Oak or Chestnut Oak, the latter name referring to the shape of the leaves, which resemble American Chestnut (pg. 169). Species name given in honor of botanist Gotthilf Henry Ernst Muhlenberg (1753–1815). While most other oaks have lobed leaves, this oak has simple leaves of variable size that tend to be gathered at the tips of branches. A member of the white oak group, producing mature acorns in one season, unlike red oaks, which take two seasons to produce mature acorns.

bark

fruit

Chestnut Oak
Quercus montana

Family: Beech (Fagaceae)

Height: 60–90' (18–27.5 m)

Tree: large tree with a straight trunk, broad open crown, often irregular

Leaf: simple, 4–8" (10–20 cm) in length, 2–4" (5–10 cm) wide, alternately attached, elliptical, rounded or pointed tip, tapering at the base, widest beyond the middle, irregular coarse-toothed margin, shiny yellow-green and almost leathery above, paler and finely hairy below

Bark: light to dark brown with deep narrow furrows

Fruit: green acorn, turning brown when mature, edible, egg-shaped (ovate), ¾–1¼" (2–3 cm) long, on a short stalk, thin cap covering upper third of nut

Fall Color: yellow to dull orange

Origin/Age: native; 200–500 years

Habitat: sandy, gravelly upland sites, hillsides, well-drained lowland sites, sun

Range: throughout

Stan's Notes: This slow-growing tree, also known as Rock Oak, is suited for planting on dry rocky soils as a shade tree. Grows on hills and mountainsides up to 4,000 feet (1,220 m). Large toothed leaves resemble chestnut tree leaves, hence its common name. Male and female flowers bloom on the same tree. Produces edible acorns each year but has a large crop every 4–7 years. Acorns are an important food crop to wildlife such as turkeys and deer. Bark contains more tannin than most other oaks. The tannin was extracted and used in tanning leather. Its strong, heavy, light brown wood is often sold as White Oak. Native to the eastern U.S. except for the deep South.

bark

fruit

Ginkgo
Ginkgo biloba

Family: Ginkgo (Ginkgoaceae)

Height: 40–60' (12–18 m)

Tree: pyramid shape, single straight trunk, narrow tapering crown

Leaf: simple, fan-shaped, 1–3" (2.5–7.5 cm) wide, alternately attached, 1 or more notches along margin, shallow irregular teeth, no midrib, veins straight and parallel (sometimes forked)

Bark: gray, irregularly rough with many furrows

Fruit: foul-smelling yellow fruit with thick, fleshy outer coat when ripe, 1" (2.5 cm) in diameter, on a long, thin fruit stalk

Fall Color: yellow

Origin/Age: non-native, introduced from eastern China; 100–150 years

Habitat: well-drained soils, sun to partial shade

Range: throughout, planted in yards and parks and along roads, not found growing in the wild

Stan's Notes: The Ginkgo is the sole surviving species from an ancient family of trees that flourished millions of years ago. Because the surviving trees were cultivated only in ancient temple gardens in China, the species remained unknown to the scientific community until the late 1700s. Only the male trees are sold and planted since female trees produce butyric acid, which makes the fruit smell foul. Ginkgo fruit has been highly prized by some people for medicinal properties. Its leaves are often in two lobes, hence the species name *biloba*. Also called Maidenhair-tree because the unique fan-shaped leaves resemble the fronds of the Maidenhair Fern plant.

flower

bark

fruit

thorn

Russian-olive

Elaeagnus angustifolia

Family: Oleaster (Elaeagnaceae)

Height: 10–20' (3–6 m)

Tree: single crooked trunk is often divided low, open irregular crown

Leaf: simple, lance-shaped, 1–4" (2.5–10 cm) in length, alternately attached, blunt tip or sharp tip, margin lacking teeth, gray above and below, leaves and twigs covered with grayish-white hairs

Bark: light gray with shallow furrows, thorns on twigs

Flower: 4-petaled yellow flower, ¼–½" (.6–1 cm) wide

Fruit: dry gray-to-yellow olive-like fruit (drupe), ¼–½" (.6–1 cm) diameter, containing 1 seed

Fall Color: brown

Origin/Age: non-native, introduced from Europe; 50–75 years

Habitat: wide variety of soils, sun to partial shade

Range: throughout, usually seen near old farmsteads, parks, formerly planted as an ornamental

Stan's Notes: Was planted in North America for its unusual gray leaves and olive-like fruit, often as a shelterbelt. While it's no longer planted, it has escaped from gardens, yards and parks and now grows in the wild (naturalized). Spread by birds, which pass the seeds through their digestive tracts unharmed. Twigs are often scaly and armed with very long thorns (see inset) that have a salmon-colored pith. The species name *angustifolia* means "narrow leaf." Also called Oleaster or Narrow-leaved Oleaster.

immature
fruit

bark

flower

fruit

Hackberry
Celtis occidentalis

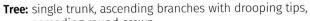

Family: Hemp (Cannabaceae)

Height: 40–60' (12–18 m)

Tree: single trunk, ascending branches with drooping tips, spreading round crown

Leaf: simple, lance-shaped, 2–4" (5–10 cm) long, alternately attached, long tapering tip, asymmetrical leaf base, evenly spaced sharp teeth, hairs on the veins, dark green above, paler below

Bark: unique gray-colored bark is covered with narrow corky ridges, wart-like

Flower: tiny green flower, ⅛" (.3 cm) wide, sprouting from bases of young leaves in early spring

Fruit: green berry-like fruit (drupe), turning deep purple at maturity, sweet and edible when ripe, ¼" (.6 cm) diameter, containing 1 seed

Fall Color: yellow

Origin/Age: native; 100–150 years

Habitat: wide variety of soils, partial shade

Range: southern half of the state

Stan's Notes: Unique corky bark makes this tree easy to identify. In fall, mature trees are laden with dark purple berry-like fruit, which typically doesn't last long because it is a favorite food of many bird species. Fruit can remain on the trees throughout winter if not eaten by birds. Commonly found in floodplain forests, but also grows well in drier areas. Suffers from non-fatal witches'-broom—dense clusters of small short twigs at branch ends caused by the combined efforts of a small insect and a fungus. Often has dimple-like galls on leaves (which do not affect the tree's health) caused by mite insects. Also called Northern Hackberry, Sugarberry or Hack-tree.

thorn

bark

flower

fruit

Hawthorn

Crataegus spp.

Family: Rose (Rosaceae)

Height: 15–25' (4.5–7.5 m)

Tree: short round tree, single trunk, flat-topped crown

Leaf: simple, oval to triangular, 2–4" (5–10 cm) in length, alternately attached, sometimes 3 lobes, double-toothed margin, thick and shiny, dark green above and below

Bark: gray with red patches, very scaly and covered with peeling bark, sturdy thorns, 1–3" (2.5–7.5 cm) in length, on branches

Flower: 5-petaled white (occasionally pink) flower, 1–2" (2.5–5 cm) diameter, in flat-topped clusters, 3–5" (7.5–12.5 cm) wide, fragrant

Fruit: red (sometimes orange) apple-like fruit (pome), edible, ½–1" (1–2.5 cm) diameter, in clusters

Fall Color: red to orange

Origin/Age: native and non-native; 50–100 years

Habitat: dry soils, open fields, hillsides, sun

Range: throughout, planted in yards and parks

Stan's Notes: One of over 100 species in North America. Because of its affinity to hybridize, there are more than 1,100 different kinds of hawthorn in the U.S., making it difficult to distinguish each. Fruit is edible and eaten by many bird and animal species. Branches and trunk are armed with long sturdy thorns that allow shrikes, also known as Butcher Birds, to impale their prey on the spines. Birds like to build their nests in this tree, gaining some protection from the large sharp thorns. Also called Thornapple or Haws.

bark

flower

fruit

Juneberry
Amelanchier arborea

Family: Rose (Rosaceae)

Height: 10–20' (3–6 m)

Tree: multiple narrow trunks, round crown

Leaf: simple, oval, 2–4" (5–10 cm) in length, alternately attached, pointed tip, fine-toothed margin (sometimes toothless near stalk), dark green above, whitish hairs below and covering stalk

Bark: light gray, smooth with shallow cracks

Flower: erect white flower, 1" (2.5 cm) long, in clusters, 1–3" (2.5–7.5 cm) wide

Fruit: red berry-like fruit (pome), turning dark blue at maturity, edible, round, ¼" (.6 cm) diameter, on a fruit stalk, in hanging clusters

Fall Color: yellow to red

Origin/Age: native; 10–20 years

Habitat: dry soils, hillsides, forest edges, open fields, sun

Range: throughout

Stan's Notes: There are up to 20 species of *Amelanchier* in the world, most occurring in North America. In Pennsylvania there are several varieties of Juneberry, but they are hard to differentiate because of crossbreeding among species. Called Juneberry because the fruit ripens in June. Fruit is edible and an important source of food for wildlife. Also known as Downy Serviceberry, "Downy" for the whitish downy hairs on the underside of leaves and covering the leafstalks, and "Serviceberry" because its flowering coincides with the time of year that cemetery burials were historically held in northern climates. Also known as Shadbush because the blooming time occurs when shad (a kind of fish) spawn.

mature fruit

bark

flower

immature fruit

Ironwood

Ostrya virginiana

Family: Birch (Betulaceae)

Height: 20–40' (6–12 m)

Tree: single trunk that is often crooked, with spreading branches, open irregular crown

Leaf: simple, oval, 2–4" (5–10 cm) in length, alternately attached, with pointed tip, asymmetrical leaf base, double-toothed margin, fuzzy to touch, yellowish-green color

Bark: gray, fibrous with narrow ridges spiraling around the trunk

Flower: catkin, ½–1" (1–2.5 cm) long

Fruit: flattened nutlet, ¼" (.6 cm) wide, within a hanging cluster of inflated sacs that are green when young, turning brown at maturity, 1½–2" (4–5 cm) long, appearing like the fruit of the hop plant

Fall Color: yellow

Origin/Age: native; 75–100 years

Habitat: dry soils, slopes, ridges, shade tolerant

Range: throughout

Stan's Notes: An important understory tree, frequently spending its entire life in the shade of other taller trees. The common name refers to its very strong and heavy wood, which is used to make tool handles and tent stakes. Distinctive thin scaly bark that spirals up the trunk and velvety soft leaves make this tree easy to identify. Also called Hop-hornbeam in direct reference to the fruit sacs, which appear like hops. One of three species of *Ostrya* in the U.S.

bark

Blue Beech

Carpinus caroliniana

Family: Birch (Betulaceae)

Height: 15–25' (4.5–7.5 m)

Tree: single to multiple crooked trunks, wide and often flat crown

Leaf: simple, oval, 2–5" (5–12.5 cm) length, alternately attached, with pointed tip, asymmetrical leaf base, margin is fine, sharp and double-toothed, color is light green

Bark: light blue-gray to gray, very smooth and unbroken with longitudinal muscle-like ridges

Fruit: many small ribbed nutlets, each ¼" (.6 cm) wide, contained in a leaf-like papery green bract, 2–4" (5–10 cm) long, that hangs in clusters and turns brown when mature

Fall Color: orange to deep red

Origin/Age: native; 50–75 years

Habitat: rich moist soils, moist valleys, along streams and other wet places, partial shade

Range: throughout

Stan's Notes: One of about 25 species of *Carpinus*, Blue Beech is the only native of North America. An easily recognized understory tree, with its smooth unbroken trunk and long, fluted muscle-like ridges. Also known as Musclewood, Water Beech, Ironwood (same common name, but not the Ironwood on pg. 141) or the American Hornbeam. In the latter name, "Horn" means "tough" and "beam" means "tree" in Old English, which describes its tough wood. The wood is used for tool handles.

bark

fruit

American Beech
Fagus grandifolia

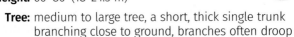

Family: Beech (Fagaceae)

Height: 60–80' (18–24.5 m)

Tree: medium to large tree, a short, thick single trunk branching close to ground, branches often droop to the ground, broad spreading round crown

Leaf: simple, oval, 2–5" (5–12.5 cm) in length, alternately attached, long pointed tip, straight parallel veins, each ends in a sharp shallow tooth, leathery dark green above, lighter green below

Bark: light gray, smooth

Fruit: reddish-brown capsule, ½–1½" (1–4 cm) in length, in pairs, splitting open into 4 sections to release a 3-sided nut

Fall Color: yellow to brown

Origin/Age: native; 150–200 years

Habitat: well-drained moist soils and bottomlands, shade tolerant

Range: throughout

Stan's Notes: Highly prized tree with important benefits to wildlife. Squirrels, grouse, bear, raccoons, deer and many other animals eat the abundant and edible beechnuts. Unusual bark in that it remains smooth even as the tree matures. The wood is very valuable and has been used for many years in furniture and flooring. One of the most abundant, well-recognized trees in eastern North America. Grows in mixed deciduous forests with oaks and maples. Can grow in pure stands. This tree has been planted in parks and around homes for many years. Until a larger tree falls, allowing enough light and room to grow, young saplings can sustain in dense shade for years.

bark

flower

fruit

Alternate-leaf Dogwood
Cornus alternifolia

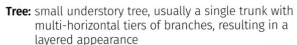

Family: Dogwood (Cornaceae)

Height: 25–35' (7.5–11 m)

Tree: small understory tree, usually a single trunk with multi-horizontal tiers of branches, resulting in a layered appearance

Leaf: simple, oval, 2–5" (5–12.5 cm) in length, alternately attached, pointed tip, margin lacking teeth, deep-curving (arcuate) parallel veins, dark green above, whitish below, leaves are often clustered at ends of the branches

Bark: gray, thin and smooth

Flower: a small white-to-cream flower, ¼–½" (.6–1 cm) wide, in round clusters, 3–5" (7.5–12.5 cm) wide

Fruit: green berry-like fruit (drupe), turning white to blue to nearly black at maturity, round, ¼" (.6 cm) diameter, on a red fruit stalk

Fall Color: red

Origin/Age: native; 40–60 years

Habitat: well-drained open woodlands, forest edges, shade

Range: throughout

Stan's Notes: A common understory tree that rarely gets taller than 35 feet (11 m), typically cultivated as an ornamental tree for landscaping. Its alternately attached leaves are unique, as other dogwoods have oppositely attached leaves. Branches are distinctly horizontal, growing in tiers, hence another common name, Pagoda Dogwood. "Pagoda" refers to a type of religious building in East Asia that has many stories or levels, just like the many levels of branches on this tree. Look for the clusters of oval leaves near ends of branches.

bark

fruit

Black Tupelo
Nyssa sylvatica

SIMPLE ALTERNATE

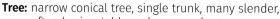

Family: Tupelo (Nyssaceae)

Height: 50–70' (15–21 m)

Tree: narrow conical tree, single trunk, many slender, often horizontal branches, round crown

Leaf: simple, oval, 2–5" (5–12.5 cm) long, alternately attached, usually widest above center of leaf, wavy toothless margin, shiny green above, paler and often hairy below, leafstalk often red, leaves often clustered at ends of branches

Bark: light gray to light brown, flaky texture with thick irregular rectangular ridges

Fruit: green berry-like fruit (drupe), turning blue-black when mature, ½" (1 cm) wide, containing 1 seed

Fall Color: red

Origin/Age: native; 50–100 years

Habitat: moist to wet soils, along streams or wetlands, partial shade

Range: throughout

Stan's Notes: A common species throughout the eastern U.S. from New England to Florida. This is a medium-sized tree, often planted as an ornamental for its red leaves in fall and its autumn-ripening fruit, which birds and mammals eat. In southern states, bees that collect nectar from its spring flowers produce Tupelo honey. Also called Black-gum, Sour-gum or Pepperidge. Reproduces by sprouts growing from its roots. The genus name is derived from the Greek word *Nysa*, the region in Greece where water nymphs were presumed to live; it originally described another *Nyssa* species that grows in swamps, the Water Tupelo (not shown). Species name *sylvatica* means "of the woods," describing its habitat.

fruit

bark

Red Mulberry
Morus rubra

Family: Mulberry (Moraceae)

Height: 20–30' (6–9 m)

Tree: single trunk, divided low, with spreading branches and dense round crown

Leaf: simple, oval to multi-lobed, 2–5" (5–12.5 cm) in length, alternately attached, with coarse-toothed margin, exudes milky sap when torn, shiny green above, hairy tufts below

Bark: gray to reddish brown with uneven furrows

Fruit: green berry (aggregate fruit), turning red to black, appearing like a raspberry, made up of many tiny 1-seeded fruit, sweet and edible, ½" (1 cm) wide

Fall Color: yellow

Origin/Age: native; 50–75 years

Habitat: moist soils, floodplains, river valleys, sun to partial shade

Range: throughout, formerly planted in parks and yards

Stan's Notes: This species produces large crops of fruit, providing an important food source for wildlife, especially birds. In summer, berries ripen to red and are delicious when black. Fruit is sweet and juicy and used in jams, jellies and pies. New trees are started when seeds pass through the digestive tracts of birds unharmed and are deposited. Its common name and the species name *rubra* refer to its mostly red fruit. Early settlers and American Indians used its fresh fruit to make beverages, cakes and preserves, and medicinally to treat dysentery and other ailments. Formerly planted in parks and yards, but fell out of favor due to the overabundance of fruit.

thorn

bark

flower

fruit

Osage-orange
Maclura pomifera

Family: Mulberry (Moraceae)

Height: 30–40' (9–12 m)

Tree: small to medium tree with a short, often crooked trunk, many spreading branches, irregular crown

Leaf: simple, oval, 3–5" (7.5–12.5 cm) long, alternately attached, long pointed tip, round to heart-shaped at base, a smooth, often wavy margin, no teeth, shiny dark green above, lighter below and lacking hairs, thorns, 1" (2.5 cm) long, at the leaf base

Bark: gray to brown with narrow forking ridges, orange inner bark where bark has been removed

Flower: clusters of green-to-cream flowers, 1" (2.5 cm) wide, on a thin fruit stalk, 2–3" (5–7.5 cm) long

Fruit: large, hard, fleshy, grainy balls, 3–5" (7.5–12.5 cm) wide, milky sap, many small brown seeds

Fall Color: yellow

Origin/Age: non-native; 100–200 years

Habitat: deep rich soils, moist soils, river valleys, sun

Range: southern half of the state, planted in parks, yards, along streets, as windbreaks and hedgerows

Stan's Notes: An ornamental tree named after Osage Indians, who made bows and clubs from the wood, and its dimpled fruit, whose skin resembles that of an orange. Also called Horse Apple, Hedge Apple or Mock Orange. Planted around homesteads before barbed wire was available, thorns (see inset) serving as fences. Wood is durable, but not used commercially. Yellow substance from roots was used to dye clothing and baskets. This was the first tree sample Lewis and Clark sent back from the Louisiana Territory in 1804. The oldest living tree is estimated to be 350 years old in Virginia.

bark

flower

fruit

Eastern Redbud
Cercis canadensis

Family: Pea or Bean (Fabaceae)

Height: 15–25' (4.5–7.5 m)

Tree: small tree, single or multiple thin trunks with low branching, horizontal branching on umbrella-like spreading crown

Leaf: simple, heart-shaped, 2–6" (5–15 cm) in length, alternately attached, pointed tip, smooth margin, shiny dark green, leafstalk swollen at the top

Bark: gray, smooth with reddish streaks, becoming scaly with age

Flower: pea-like lavender-to-pink flower, ¼" (.6 cm) wide, along the branches

Fruit: reddish-brown pod, 2–4" (5–10 cm) long, pointed at both ends, on a short stalk

Fall Color: yellow

Origin/Age: native; 50–75 years

Habitat: moist soils, along streams, forest edges, shade

Range: throughout, planted in parks and gardens

Stan's Notes: Also known as American Redbud or Judas-tree, the latter name referring to a legend that Judas hanged himself from a redbud tree, and that the once-white flowers are now forever red with shame. An understory tree that tolerates shade, it would not be spring in many places without its spectacular display of pink flowers, which appear before the leaves. One of two *Cercis* species native to North America. The second species is found in California, Arizona and Nevada.

immature fruit

bark

flower

fruit

Common Persimmon
Diospyros virginiana

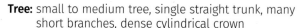

SIMPLE ALTERNATE

Family: Ebony (Ebenaceae)

Height: 30–40' (9–12 m)

Tree: small to medium tree, single straight trunk, many short branches, dense cylindrical crown

Leaf: simple, oval to elliptical, 2–6" (5–15 cm) in length, alternately attached, with pointed tip, round base, smooth margin, no teeth, shiny dark green above, whitish green below

Bark: brown to nearly black, thick and deeply furrowed into small squares

Flower: tiny white-to-cream bell-shaped, 4-lobed flower ½–¾" (1–2 cm) long, solitary at base of leafstalk

Fruit: fleshy, smooth-skinned orange-to-purplish-brown berry, wrinkled when mature, round, ¾–1½" (2–4 cm) wide, short stalk, many flat brown seeds

Fall Color: yellow

Origin/Age: native; 100–200 years

Habitat: wide variety of soils, river valleys, dry upland sites

Range: southern half of the state

Stan's Notes: Popular native tree, often planted for its glossy leaves and attractive square-patterned bark. Not easily transplanted due to its large taproot. Known for its large edible fruit, which ripens in fall and is a source of food for mammals and birds. Ripe fruit tastes like dates; used in puddings, cakes and bread. Not the same fruit sold in stores, but closely related. Often in mixed forests. Shade tolerant and slow growing, it can live in understory for many years. Close-grained, hard and strong wood is used for billiard cues. One of two native members of the Ebony family in the U.S., which includes more than 450 species around the world, mostly in tropical areas. Also called Simmon or Possumwood.

bark

flower

fruit

Witch-hazel
Hamamelis virginiana

Family: Witch-hazel (Hamamelidaceae)

Height: 20–30' (6–9 m)

Tree: small tree, multiple thin trunks are often crooked, spreading branches, broad open crown

Leaf: simple, oval to round, 3–6" (7.5–15 cm) in length, alternately attached, with pointed or rounded tip, asymmetrical base, irregular wavy margin, coarse teeth, dark green above, slightly lighter below

Bark: light brown, smooth, uniform, sometimes scaly

Flower: bright yellow flower, 4 long, thin, twisted petals, in clusters of 3, located at joint where leaf attaches to twig, blooms in autumn

Fruit: green capsule, turning orange and splitting open at maturity (autumn of the following year), ¾–1" (2–2.5 cm) long, has 2 small shiny black seeds

Fall Color: yellow

Origin/Age: native; 30–50 years

Habitat: moist soils, understory of deciduous forests, shade

Range: throughout

Stan's Notes: The name "Witch" in Witch Hazel doesn't refer to witchcraft, but comes from an old English word "Wice" referring to a plant with bendable branches. "Hazel" comes from the leaves, which resemble those of the hazel shrub (*Corylus* spp.). Leaves, twigs and bark are aromatic and have been used as an astringent in medicinal washes. This species has an unusual flowering time: in the fall. Fruit develops the following summer and matures in autumn. Capsules remain on the tree for several years, even after the seeds drop to the ground.

bark

flower

fruit

American Basswood
Tilia americana

Family: Linden (Malvaceae)

Height: 50–70' (15–21 m)

Tree: tall tree, single or multiple trunks from a common point on the ground, full round crown

Leaf: simple, heart-shaped, 3–7" (7.5–18 cm) in length, alternately attached, with asymmetrical leaf base, sharp-toothed margin, dull green above, lighter green below

Bark: light gray color and smooth when young, darkens with long, narrow, flat-topped ridges dividing into a short block with age, inner bark fibrous

Flower: creamy yellow flower, 1–2" (2.5–5 cm) diameter, fragrant scent

Fruit: nut-like green fruit, turning yellow when mature, round, ¼" (.6 cm) diameter, covered with light brown hairs, on a 1–2" (2.5–5 cm) long fruit stalk, hanging in clusters from a leaf-like wing

Fall Color: yellow, orange

Origin/Age: native; 150–200 years

Habitat: moist soils, partial shade to sun

Range: throughout

Stan's Notes: A long-lived, fast-growing tree that is well known for growing several trunks from the base of the mother plant. The light-weight soft wood is used for carving due to its smooth grain. From its flowers, bees produce a high quality honey. Fibrous inner bark was once used by American Indians to weave mats, rope and baskets. Also called American Linden or Basswood. White Basswood, once considered its own species, is now believed to be a variety of the American Basswood. Many cultivated varieties exist.

bark

fruit

cross section

Sycamore
Platanus occidentalis

Family: Sycamore (Platanaceae)

Height: 60–90' (18–27.5 m)

Tree: large tree, often a single massive trunk, enlarged at the base, open, widely spreading crown

Leaf: simple, triangular, 4–8" (10–20 cm) in length, alternately attached, 3–5 shallow pointed lobes, wavy coarse-toothed margin, 3 prominent veins, bright green above, paler below

Bark: pale white color, smooth, peeling off in large thin sections, green and cream-to-white inner bark produces a mottled effect, bark on the base of tree often much darker than upper bark

Fruit: light brown round aggregate of many nutlets, 1" (2.5 cm) in diameter, hanging from a long fruit stalk, remaining on tree into winter

Fall Color: brown

Origin/Age: native; 200–250 years

Habitat: moist soils, rich bottomlands, sun to partial shade

Range: throughout

Stan's Notes: One of the largest broadleaf trees in the state. Produces a massive white trunk, larger in diameter than many other trees in Pennsylvania, which makes it easy to identify. Its branches are often crooked, making this species easy to climb. Some hollow trunks of large old trees are used by many animals and birds as homes. It is a fast-growing tree that usually grows in old fields or along streams. Often planted as an ornamental tree in landscapes or parks. Also called American Sycamore, American Plane Tree or Buttonball-tree. Three of the 10 sycamore species grow in the U.S.; one of those grows in Canada as well.

bark

flower

fruit

Cucumbertree
Magnolia acuminata

Family: Magnolia (Magnoliaceae)

Height: 40–60' (12–18 m)

Tree: medium-sized tree, straight trunk, short, upright spreading branches, narrow pointed crown

Leaf: simple, oval, 5–10" (12.5–25 cm) long (sometimes larger), alternately attached, pointed tip, round at the base, wavy or smooth toothless margin, green above, paler below with soft fine hairs

Bark: brown with narrow furrowed rows

Flower: large 6-petaled white-to-yellow flower, 2–3½" (5–9 cm) wide, solitary at the end of branch in spring

Fruit: green aggregate, turning dark red when mature, cone-shaped, 1½–2½" (4–6 cm) long, on a long fruit stalk at the end of branch

Fall Color: dull yellow or brown

Origin/Age: native; 125–150 years

Habitat: wide variety of soils from moist to dry, usually in moist valleys, partial shade

Range: western half of the state

Stan's Notes: This species is the largest and hardiest of our native magnolias. Originally widespread and abundant. Now also planted widely as an ornamental because it is fast growing and long-lived. Takes 25–30 years to reach flowering size. Produces flowers and fruit each year, with large seed crops every 3–5 years. Seeds are eaten by many bird and mammal species. Often grows large enough to be harvested for lumber. Common name comes from its unripened fruit, which looks like a cucumber.

bark

flower

fruit

Pawpaw
Asimina triloba

Family: Custard-apple (Annonaceae)

Height: 20–30' (6–9 m)

Tree: understory tree, single or multiple straight trunks, straight branches, broad crown

Leaf: simple, wider toward the end of the leaf, 7–10" (18–25 cm) long, alternately attached, widest near the tip, smooth toothless margin, medium green above, paler below, short leafstalk (petiole)

Bark: brown, smooth, thin, covered with tiny bumps

Flower: triangular flower, 1–2" (2.5–5 cm) wide, made of 6 reddish-purple petals, solitary or in small clusters, nodding on stalk, unpleasant odor

Fruit: green berry-like fruit, turning yellow to brown or black at maturity, soft edible flesh with prune-like texture and fruity custard flavor, round to slightly curved, 3–5" (7.5–12.5 cm) long, single or in small clusters, several large seeds, ½" (1 cm) wide

Fall Color: yellow

Origin/Age: native; 100–150 years

Habitat: moist soils, floodplains, shade

Range: southern half of the state

Stan's Notes: An unusual tree with large leaves and tiny banana-shaped fruit. Also known as Wild Banana. In a mostly tropical tree family that produces fruit such as soursops and custard apples. Doesn't grow to its full potential in Pennsylvania, where it is near its northernmost growing range. New shoots grow from roots, forming large dense thickets or colonies. Flowers appear with leaves. Fruit was eaten by American Indians and early settlers and is sold in some stores today.

bark

flower

fruit

American Chestnut
Castanea dentata

SIMPLE
ALTERNATE

Family: Beech (Fagaceae)

Height: 60–90' (18–27.5 m)

Tree: once a tall tree growing to 115' (35 m), but now rarely reaching maturity due to chestnut blight, full round crown

Leaf: simple, narrow, 6–12" (15–30 cm) in length, alternately attached, tapering at each end, with unique prominent teeth that extend beyond leaf margin and form a forward curve like the teeth of a saw, straight and parallel veins

Bark: dark brown to red, smooth, separating into wide flat-topped ridges

Flower: catkin with male and female flowers on same tree (monoecious), male flower in catkin, 1" (2.5 cm) long, solitary female flower, ½" (1 cm) wide, located at base of the catkin

Fruit: bur-like green nut, turning brown at maturity, 2–3" (5–7.5 cm) wide, in clusters, splits into 4 parts

Fall Color: yellow

Origin/Age: native; 25–50 years

Habitat: well-drained sandy soils, sun

Range: throughout, often seen at old dwellings, in parks

Stan's Notes: There are five chestnut species in North America, with this one producing the proverbial chestnuts roasting over an open fire. Once a large and prominent tree of the eastern U.S., it was nearly wiped out by chestnut blight, a fungus. Grows to half its former size before dying. Hard oak-like wood with a straight grain, relatively decay resistant. Will sprout from stumps of dead or cut trees. The species name *dentata* refers to the obvious large teeth on leaves.

bark

flower

fruit

Amur Maple
Acer ginnala

Family: Soapberry (Sapindaceae)

Height: 15–20' (4.5–6 m)

Tree: small tree, trunk often multi-stemmed, compact lower branches, irregular crown

Leaf: lobed, arrowhead-shaped, 2–4" (5–10 cm) in length, oppositely attached, with 3–5 sharp lobes, coarse teeth, glossy green, leafstalk often bright red

Bark: gray, smooth with many vertical cracks

Flower: small green flower, ½" (1 cm) wide, in drooping clusters, 1–2" (2.5–5 cm) wide

Fruit: pair of red-tinged winged seeds (samara), turning bright red, 1–2" (2.5–5 cm) long

Fall Color: red to orange

Origin/Age: non-native, introduced to the U.S. from eastern Asia; 25–50 years

Habitat: well-drained soils, shade

Range: scattered throughout, planted in parks and yards

Stan's Notes: A remarkably attractive small tree, widely planted as an ornamental shrub for its brilliance in autumn. Very hardy, shade-tolerant plant that turns a showy bright red each fall, with winged seeds (samara) also turning bright red. The samaras are often called helicopters due to the way they rotate to the ground. The genus name *Acer* is Latin and means "sharp," referring to its pointed lobes. The common name comes from the Amur River, which forms the boundary between China and Russia, presumably the origin of this species. Also called Siberian Maple.

bark

flower

fruit

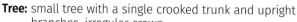

Mountain Maple
Acer spicatum

Family: Soapberry (Sapindaceae)

Height: 20–30' (6–9 m)

Tree: small tree with a single crooked trunk and upright branches, irregular crown

Leaf: lobed, 2–4" (5–10 cm) long, oppositely attached, 3 pointed lobes (rarely 5), coarse-toothed margin, light green, leafstalk often red, usually longer than the leaf

Bark: reddish to brown, smooth but grooved with light-colored areas

Flower: many 5-petaled yellowish-green flowers, each ½" (1 cm) diameter, growing in spike clusters, 1–3" (2.5–7.5 cm) tall

Fruit: pair of winged seeds (samara), often red, turning yellow, then brown, ¾–1" (2–2.5 cm) long

Fall Color: red or orange

Origin/Age: native; 40–60 years

Habitat: moist soils, along streams and other wet areas, shade tolerant

Range: throughout

Stan's Notes: Usually is considered an understory tree because it commonly grows under the canopy of larger, more dominant trees. Doesn't do well in the open. Often grows in rocky outcroppings. Also called Moose Maple since it frequently grows in habitats that are good for moose. The winged seeds are often bright red, later turning yellow, then brown in late summer before falling in early winter. Shallow root system. The wood hasn't been considered for any commercial use.

bark

flower

fruit

Red Maple
Acer rubrum

Family: Soapberry (Sapindaceae)

Height: 40–60' (12–18 m)

Tree: single trunk, narrow dense crown

Leaf: lobed, 3–4" (7.5–10 cm) long, oppositely attached, 3–5 lobes (usually 3), shallow notches in between lobes, double-toothed margin, light green color, red leafstalk

Bark: gray, smooth, broken by narrow irregular cracks

Flower: tiny red hanging flower, ¼" (.6 cm) wide, on a 1–2" (2.5–5 cm) long red stalk, growing in clusters, 1–3" (2.5–7.5 cm) wide

Fruit: pair of winged seeds (samara), red in springtime, ½–1" (1–2.5 cm) long

Fall Color: red to orange

Origin/Age: native; 75–100 years

Habitat: wet to moist soils, along swamps or depressions that hold water, sun to partial shade

Range: throughout

Stan's Notes: One of the most drought-tolerant species of maple in the state. Often planted as an ornamental, it can be identified by its characteristic leaves, which have three pointed lobes and red stalks. The common name comes from the obvious red flowers that bloom early in spring, but the flowers and leafstalks are not the only red colors it has. New leaves, fall color and spring seeds are also red. Produces one of the smallest seeds of any of the maples. Also called Swamp Maple, Water Maple or Soft Maple, the latter being used to refer to the Silver Maple (pg. 181) as well. Even though it is sometimes called Soft Maple, its wood is very hard and brittle.

bark

flower

fruit

Sugar Maple
Acer saccharum

Family: Soapberry (Sapindaceae)

Height: 50–70' (15–21 m)

Tree: single trunk, ascending branches, narrow round to oval crown

Leaf: lobed, 3–5" (7.5–12.5 cm) in length, oppositely attached, 5 lobes (occasionally 3), pointed tips, few irregular teeth, wavy margin, yellowish green above, paler below

Bark: gray in color, narrow furrows and irregular ridges, can be scaly

Flower: greenish-yellow flower, ¼" (.6 cm) wide, dangling on a 1–2" (2.5–5 cm) long stalk

Fruit: pair of green winged seeds (samara), turning tan, ¾–1½" (2–4 cm) long

Fall Color: orange to red

Origin/Age: native; 150–200 years

Habitat: rich moist soils, sun

Range: throughout, planted in yards, parks and along roads

Stan's Notes: A popular and well-known tree, Sugar Maples are the source of maple syrup and maple sugar. It takes approximately 40 gallons of sap to make a single gallon of syrup. Any break in a twig, branch or trunk leaks sugary water in spring, attracting birds, bugs and mammals, which lap up the sap. Leaves that have fallen break down quickly, making it one of the best natural and organic fertilizers. Also known as Hard Maple, its extremely hard wood is used in furniture, flooring and cabinets. Often planted in yards and parks and along roads. This species closely resembles the Black Maple (pg. 179).

bark

fruit

Black Maple

Acer nigrum

Family: Soapberry (Sapindaceae)

Height: 40–70' (12–21 m)

Tree: medium tree, nearly identical to the Sugar Maple, broad round crown

Leaf: lobed, 3–6" (7.5–15 cm) long, oppositely attached, 3 pointed lobes (rarely 5), with a smooth to wavy margin, dark green above, yellowish below, leaves appear to droop

Bark: light gray, smooth, can get scaly with age

Fruit: pair of green winged seeds (samara), turning tan, 1–2" (2.5–5 cm) long

Fall Color: yellow

Origin/Age: native; 150–175 years

Habitat: moist fertile soils, floodplains, bottomlands, shade

Range: western quarter of the state

Stan's Notes: Nearly identical to the Sugar Maple (pg. 177), with several areas of differentiation. Common name "Black" refers to its bark, which is not black but often is darker than Sugar Maple bark. Leaves occasionally have dense, velvety brownish hairs underneath. They sometimes also have a characteristic wilted appearance and turn yellow in fall rather than red, like Sugar Maple leaves. Black Maple trees grow in moister soils and are more drought tolerant and slower growing than Sugar Maples, but they crossbreed easily with each other. Seeds lying in soil remain viable for many years. Because Black and Sugar Maples are so similar, some think they should be considered subspecies, not separate species.

underside

bark

flower

fruit

Silver Maple
Acer saccharinum

Family: Soapberry (Sapindaceae)

Height: 75–100' (23–30.5 m)

Tree: single trunk, ascending branches, open crown

Leaf: lobed, 4–6" (10–15 cm) long, oppositely attached, 5–7 lobes, pointed tips, deep notches and double-toothed margin, dull green above with a silvery white color below

Bark: gray and smooth when young, becomes furrowed, long scaly strips, often peeling and curling at ends

Flower: tiny red dangling flower, ¼" (.6 cm) wide, on 1–2" (2.5–5 cm) long stalk

Fruit: pair of green winged seeds (samara), turning to brown, 1–2½" (2.5–6 cm) long

Fall Color: yellow to orange

Origin/Age: native; 100–125 years

Habitat: wet to moist soils, often growing in pure stands in floodplains, shade

Range: throughout

Stan's Notes: Usually seen growing in bottomlands or floodplains along rivers, where it is often the dominant tree. This is one of the first trees to bloom (flower) in spring, confusing many to think it is budding early. Bark of older trees is characteristic, with long strips that often peel and curl at ends. Produces heavy seed crops. Also called Silver-leaf Maple for the silvery appearance of the underside of leaves, and Soft Maple, which is also another name of the Red Maple (pg. 175). This name refers to the brittle branches (which often break off in windstorms), rather than to the wood being soft. The wood is actually very hard with a tight grain.

fruit

bark

flower

Norway Maple
Acer platanoides

Family: Soapberry (Sapindaceae)

Height: 40–60' (12–18 m)

Tree: single straight trunk, dense round crown

Leaf: lobed, 5–7" (12.5–18 cm) in length, oppositely attached, 5–7 lobes, shallow notches and a wavy margin, exudes milky sap when cut, shiny dark green above, light green below

Bark: dark gray in color with many narrow furrows and interlacing ridges

Flower: large green flower, ½–¾" (1–2 cm) wide, on a 1–2" (2.5–5 cm) long green stalk

Fruit: pair of widely spread winged seeds (samara), 1–2" (2.5–5 cm) long

Fall Color: yellow to orange

Origin/Age: non-native, introduced to the U.S. from Europe; 100–125 years

Habitat: well-drained rich soils, sun to partial shade

Range: throughout, planted in parks, yards and along streets

Stan's Notes: This introduced species, most commonly seen along streets and in parks, has spread to wild environments and is doing well. Considered to be one of the most disease- and insect-resistant species of maple and a potential species that could outperform the widely prevalent native maples. While leaves are similar to those of Sugar Maple (pg. 177), several Norway Maple varieties have red or purple leaves. Leaves, buds and twigs exude a milky sap when cut. Winged seeds are more widely spread than those of Sugar Maple or Silver Maple (pg. 181). Common name implies it was introduced from Norway.

bark

flower

fruit

Striped Maple
Acer pensylvanicum

Family: Soapberry (Sapindaceae)

Height: 10–30' (3–9 m)

Tree: single or multiple crooked trunks, many short thin branches, open conical or flat crown

Leaf: lobed, 5–7" (12.5–18 cm) in length, oppositely attached, with 3 short pointed lobes near the tip, occasionally without lobes, sharp-toothed margin, 3 main veins extend from the leaf base to the end of each lobe, light green above, paler below, stout leafstalk, 2–2½" (5–6 cm) long

Bark: green with vertical white stripes, becoming darker with age

Flower: many 5-petaled, bell-shaped yellow flowers, ⅜" (.9 cm) long, on a long stalk, hanging vertically

Fruit: pair of winged seeds (samara), 1¼" (3 cm) long

Fall Color: yellow

Origin/Age: native; 50–75 years

Habitat: moist upland soils, understory of forests, shade

Range: throughout, except for southeastern Pennsylvania

Stan's Notes: A small ornamental tree, often confused with a shrub. Easy to identify, even in winter, by its unique striped bark. If pruned it will develop with a single trunk, but if left unchecked it will grow with many trunks. Grows in the shade of other larger trees in the forest. White-tailed Deer browse on the leaves in spring and summer and twigs and saplings in winter. Also called Moosewood, presumably because moose like to browse on its twigs and leaves.

underside

bark

flower

fruit

White Poplar
Populus alba

Family: Willow (Salicaceae)

Height: 40–60' (12–18 m)

Tree: medium-sized tree with single or multiple trunks, open, widely spreading crown

Leaf: lobed, maple-shaped, 2–5" (5–12.5 cm) in length, alternately attached, 3 pointed lobes, few rounded teeth, light green above and chalky white below, covered with white hairs, silky white when young

Bark: dark brown color and deeply furrowed near base, yellowish-white color with dark horizontal marks (lenticels) and smooth upper

Flower: catkin, 2–3" (5–7.5 cm) long, composed of many tiny flowers, ¼" (.6 cm) wide

Fruit: catkin-like fruit, 2–3" (5–7.5 cm) long, composed of many capsules that open and release many tiny cottony seeds, which float on the wind

Fall Color: yellow to brown

Origin/Age: non-native, introduced to the U.S. from Europe; 100–125 years

Habitat: wide variety of soils, sun

Range: throughout, planted in parks, yards and along roads

Stan's Notes: The maple-like lobed leaves of the White Poplar are unusual for a member of the *Populus* genus. Buds and leaf undersides are covered with tiny white hairs, giving newly budded leaves a whitish-colored appearance and the species its common name. Also known as Silver-leaf Poplar or European White Poplar, it was among the first trees that were introduced to North America from Europe during colonial times. A fast-growing tree with several varieties sold. Species name *alba* means "white."

bark

flower

fruit

Sassafras
Sassafras albidum

Family: Laurel (Lauraceae)

Height: 30–60' (9–18 m)

Tree: medium-sized columnar tree with single crooked trunk, branches often crooked and spreading, flat irregular crown

Leaf: lobed or simple, 3–5" (7.5–12.5 cm) in length, alternately attached, often has 1, 2 or 3 lobes with mitten-shaped 1-lobed leaves, simple elliptical leaves along with lobed leaves appear on the same tree, smooth toothless margin, shiny green above, paler below and often hairy

Bark: brown, deeply furrowed with age

Flower: green-to-yellow flower, ¼" (.6 cm) wide, on a 1–2" (2.5–5 cm) long stalk, in clusters

Fruit: blue-to-nearly-black fruit (drupe), ½" (1 cm) wide, on a 1" (2.5 cm) long fruit stalk, in clusters, each containing 1 seed

Fall Color: yellow to red

Origin/Age: native; 100–150 years

Habitat: moist and dry soils, forest edges, sun to partial sun

Range: throughout

Stan's Notes: Like other laurels, crushed leaves and twigs have a spicy fragrance. Its shallow roots once supplied oil of sassafras for perfuming soap. Once thought to have medicinal properties, the settlers brewed its bark and roots into a tea. A compound in Sassafras is suspected to be a carcinogen as is banned as a food additive. One of our first exports. Has an American Indian name. Produces dense cover and fruit for wildlife. Sometimes planted as an ornamental for its fall colors and fruit. Soft wood breaks easily.

flower

bark

immature fruit

fruit

Sweetgum
Liquidambar styraciflua

Family: Sweetgum (Altingiaceae)

Height: 80–100' (24.5–30.5 m)

Tree: tall tree, single straight trunk, wide at the bottom, narrow at the top, pointed crown

Leaf: lobed, 3–6" (7.5–15 cm) long, alternately attached, star-shaped, 5–7 long pointed lobes, fine-toothed, 5 main veins from a notched base to tip of each lobe, shiny dark green above, whitish green below

Bark: gray to brown, deeply furrowed into narrow scales

Flower: ball-shaped green flower, ½–¾" (1–2 cm) wide, in clusters, on a 1–2" (2.5–5 cm) drooping stalk

Fruit: round green cluster, 1–1½" (2.5–4 cm) diameter, turns brown at maturity, many hard woody spines, on a long fruit stalk, has many flat, winged seeds

Fall Color: red

Origin/Age: native; 100–150 years

Habitat: moist soils, river valleys, mixed woods, old fields

Range: southern half of the state

Stan's Notes: Often planted as a shade tree and for its bright red leaves in fall. Can be in pure stands, but usually grows with other deciduous trees. Moderately rapid growing. Matures at 20–25 years. Flowers in spring. Fruit matures in fall and opens to release seeds. Seeds eaten by birds and animals. Old fruit may stay on the tree in winter. Is an important tree commercially, surpassed only by oaks. Dark reddish-brown, hard, heavy wood is used for furniture veneer, plywood and barrels. Well known for its resin, obtained by peeling back bark and scraping off resin. The gum has been used medicinally, in soaps and in adhesives and for chewing gum. The leaves are aromatic if crushed.

bark

flower

fruit

Tulip-tree
Liriodendron tulipifera

Family: Magnolia (Magnoliaceae)

Height: 80–100' (24.5–30.5 m)

Tree: large tree, single straight trunk, branch-free lower half, ascending upper branches, drooping lower branches, narrow crown

Leaf: lobed, 3–6" (7.5–15 cm) long, alternately attached, unusual shapes, some nearly square but most with 4–6 pointed lobes, shiny green above, paler below, smooth toothless margin, very long leafstalk

Bark: gray, deeply furrowed with age

Flower: large, showy, tulip-shaped flower, 1–2" (2.5–5 cm) long, made of 6 yellow petals with orange bases

Fruit: green aggregate, turning light brown at maturity, cone-shaped, 3" (7.5 cm) long, growing upright on branch, containing tiny winged nutlets that fall apart in autumn and release seeds

Fall Color: yellow

Origin/Age: native; 100–150 years

Habitat: moist soils, along streams, ponds and other wet places, sun

Range: throughout

Stan's Notes: Often one of the tallest trees in the forest. Also called Yellow Poplar due to the heartwood color, but it is not a poplar. Two *Liriodendron* species, one in China. Fossil evidence shows these trees were once found throughout much more of the northern hemisphere. Its colorful flowers give it the common name. Fast growing, with some reaching nearly 200 feet (61 m). Loses its lower leaves as it develops. A tonic from its bitter, aromatic roots was used for heart ailments.

bark

fruit

English Oak

Quercus robur

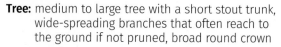

Family: Beech (Fagaceae)

Height: 60-80' (18-24.5 m)

Tree: medium to large tree with a short stout trunk, wide-spreading branches that often reach to the ground if not pruned, broad round crown

Leaf: lobed, oblong, 2–6" (5–15 cm) in length, alternately attached, many shallow lobes, widest at the top, tapers at the base, very short leafstalk, dark green above, paler below

Bark: dark gray with deep irregular furrows

Fruit: green acorn, turning brown when mature, edible, egg-shaped, ¾–1" (2–2.5 cm) long, 1–5 acorns in a cluster, each on a long stalk, thin cap covering the upper third of nut

Fall Color: yellow to brown

Origin/Age: non-native; 200–250 years

Habitat: wide variety of soils, forest edges

Range: throughout, planted in parks, yards, along roads

Stan's Notes: This non-native species has been planted in North America since colonial times. Greatest densities are in eastern states, where it was introduced. Its rapid growth, good form and dense shade make it a desirable tree for planting in parks, yards and along streets. This tree is long-lived, cold tolerant and transplants easily. Also called Pedunculate Oak because its acorns grow on long stalks, or peduncles. Acorns are eaten and transported by many bird and animal species, enabling it to escape cultivation. Now grows in the wild. In Europe the wood was used for building ships and as paneling for important buildings. Some trees in England are reported to be more than 1,000 years old.

bark

fruit

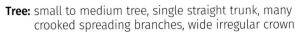
Blackjack Oak

Quercus marilandica

Family: Beech (Fagaceae)

Height: 25–40' (7.5–12 m)

Tree: small to medium tree, single straight trunk, many crooked spreading branches, wide irregular crown

Leaf: lobed, 2–6" (5–15 cm) long, alternately attached, 3–5 broad, shallowly divided lobes, lobes broadest at the top, each lobe bristle-tipped, leathery, waxy, shiny dark green above and dull green below with brown hairs along veins

Bark: nearly black bark that is rough, thick and deeply furrowed into nearly square plates

Fruit: green acorn, turning brown when mature, edible, oblong, ½–¾" (1–2 cm) long, with a stout pointed tip, single or in pairs on a very short stalk, cap covers the upper half of nut, matures in 2 seasons

Fall Color: brown to yellow

Origin/Age: native; 150–200 years

Habitat: dry, sandy or clay soils, often on upland ridges, sun

Range: scattered in far southeastern corner of the state

Stan's Notes: A red oak of the deep South, reaching its northern limits in Pennsylvania. Frequently associated with poor soils. Tolerates dry, sandy or gravelly soils where few other forest oaks live. Also called Scrub Oak due to its stunted, scrubby growth. First described in the 1770s from a tree found in Maryland, hence the species name. Club-shaped leaves give it the common name. Leaves are covered with a waxy substance (cutin) that helps reduce water loss in hot, dry environments. Retains leaves well into winter. Small acorns take two years to mature. New shoots sprout from burned or cut trunks. Wood is sometimes used for railroad ties and firewood.

bark

fruit

Pin Oak
Quercus palustris

Family: Beech (Fagaceae)

Height: 50–70' (15–21 m)

Tree: medium-sized tree with a single trunk, drooping branches and open irregular crown

Leaf: lobed, 3–5" (7.5–12.5 cm) in length, alternately attached, 5–7 lobes, each ending in a pointed tip (bristle-tipped), deep sinuses in between the lobes cutting nearly to the midrib, shiny green above, light green below with tufts of hair along midrib

Bark: dark brown with shallow furrows and flat ridges

Flower: green-to-red catkin, 1–4" (2.5–10 cm) long, made up of many tiny flowers, ⅛" (.3 cm) wide

Fruit: green acorn, turning brown when mature, edible, ½" (1 cm) long, elongated cap covering the upper third of nut, maturing at the end of the second year

Fall Color: deep red to reddish brown

Origin/Age: native; 100–150 years

Habitat: wet soils, frequently in heavy clay, poorly drained upland sites, sun

Range: southern half of the state

Stan's Notes: One of over 600 species of oak in the world. Often grows in pure stands. Flowers are pollinated in the first year; acorns mature in fall of the second year. Heavy fruit crops every 4–6 years. Acorns are eaten by wildlife. Acorns contain tannic acid (tannin), which can be toxic in large amounts. Nuts should be processed by boiling in several changes of water before eating. Susceptible to oak wilt, a fungus that can kill entire stands of trees. Northern Pin Oak (*Q. ellipsoidalis*) (not shown) of the upper Midwest has similar leaves. Also called Spanish Oak.

bark

fruit

Post Oak
Quercus stellata

Family: Beech (Fagaceae)

Height: 30–60' (9–18 m)

Tree: stout tree with a straight trunk and wide-spreading branches, round crown

Leaf: lobed, unique shape suggests a Maltese cross, 3–7" (7.5–18 cm) long, alternately attached, 5–7 deep, squared lobes with 2 large middle lobes, rounded tip, tapered base, dark green above, paler below, leafstalk often stout and hairy, yellow star-shaped hairs on underside of leaf

Bark: light gray with scaly ridges

Fruit: green acorn, turning brown when mature, edible, egg-shaped (ovate), ½–1" (1–2.5 cm) long, single on a very short stalk or stalkless, thin cap covering the upper third of nut

Fall Color: brown

Origin/Age: native; 200–250 years

Habitat: sandy, gravelly soils, along hills and ridges, along streams and rivers in floodplains, sun

Range: southeastern quarter of the state

Stan's Notes: Medium-sized oak that reaches its northern growing limits in Pennsylvania. Slow-growing, drought resistant, often planted in dry rocky soils where other oaks won't survive. Frequently scrubby due to nutrient-poor soils. Unique leaf shape makes it easy to identify, but shape varies. Some leaves look like a cross. Thick yellowish twigs and star-shaped hairs on leaves help to identify. *Stellata* refers to the star shape of the leaf hairs. Wood was used for fence posts, hence the common name. Produces acorns each year, but has heavy crops every 2–4 years. An important food source for wildlife.

bark

fruit

Scarlet Oak
Quercus coccinea

Family: Beech (Fagaceae)

Height: 60–70' (18–21 m)

Tree: medium tree with a straight trunk, wide-spreading branches, round crown

Leaf: lobed, 3–7" (7.5–18 cm) long, alternately attached, 5–7 pointed lobes, round sinuses, deeply divided nearly to the midrib, dark green and glossy above, paler yellowish below, leafstalk often reddish

Bark: dark gray, becoming darker with age, scaly ridges

Fruit: green acorn, turning brown when mature, edible, egg-shaped (ovate), ½–1" (1–2.5 cm) long, single on a short stalk, cap covering the upper third to half of nut, maturing in 2 seasons

Fall Color: scarlet red

Origin/Age: native; 100–150 years

Habitat: dry, sandy or gravelly soils, along hills and ridges, often with other oaks, sun

Range: southern half of the state, planted in parks, yards and along roads

Stan's Notes: A component of eastern deciduous forests, growing with many other oaks, such as White Oak (pg. 209) and Black Oak (pg. 211). Planted in parks and yards and along streets for its form, exceptional shade and brilliant fall color, for which it was named. Drought tolerant and fast growing, this tree thrives in well-drained soils. Not as long-lived as many other oaks. A red oak, producing flowers in the spring. Acorns mature at the end of the second year. Heavy acorn crops are often followed by 3–4 years with few or no acorns. Lumber is marketed as Red Oak but is not as desirable as White Oak wood.

bark

fruit

Shumard Oak
Quercus shumardii

Family: Beech (Fagaceae)

Height: 60–90' (18–27.5 m)

Tree: large tree, single straight trunk, wide round crown

Leaf: lobed, 3–7" (7.5–18 cm) long, alternately attached, 5–9 pointed lobes, round sinuses, deeply divided nearly to the midrib, broadest at the top, several bristle-tipped teeth, dark green above, dull green below with tufts of hair at vein angles

Bark: gray, becoming darker with age, shallow furrows

Fruit: green acorn, turning brown when mature, edible, egg-shaped (ovate), ¾–1¼" (2–3 cm) long, single or in pairs on a short stalk, cap covering the upper third to half of nut, maturing in 2 seasons

Fall Color: red to brown

Origin/Age: native; 150–200 years

Habitat: moist soils along rivers, streams and swamps or scattered in deciduous forests and river valleys

Range: scattered in extreme southern Pennsylvania

Stan's Notes: A larger oak in North America, some growing nearly 100 feet (30.5 m) tall, with a very wide crown. Grows the best in well-drained soils along rivers. Usually is considered an oak of the Deep South, where it is planted as an ornamental tree. Hybridizes with other oak species, especially Shingle Oak (pg. 125) and Blackjack Oak (pg. 197), making them hard to distinguish. Often substituted for Scarlet Oak (pg. 203) in landscaping. Moderately fast-growing tree that flowers in spring. Acorns mature in the fall of its second year. Very large acorn crops every 4–6 years. Wood has been used for flooring, veneers and furniture. Sold as Red Oak. Named for B. F. Shumard, state geologist of Texas in the mid-1800s.

bark

flower

fruit

Swamp White Oak

Quercus bicolor

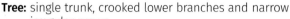

Family: Beech (Fagaceae)

Height: 40–60' (12–18 m)

Tree: single trunk, crooked lower branches and narrow irregular crown

Leaf: lobed, 4–7" (10–18 cm) long, alternately attached, widest above the middle, with shallow lobes that occasionally appear like teeth, dark green above, paler below with white hairs, obvious difference between upper and lower leaf surfaces

Bark: light gray, many vertical furrows, wide flat ridges

Flower: thin catkin, 1–4" (2.5–10 cm) long, composed of hairy green flowers

Fruit: green acorn, turning brown when mature, edible, ¾–1¼" (2–3 cm) long, single but can be in pairs, knobby cap covering the upper half of nut

Fall Color: brown

Origin/Age: native; 150–200 years

Habitat: moist soils, along river bottoms, wetland edges

Range: southern half of the state

Stan's Notes: This is a common, fast-growing oak of moist soils. Its acorns, which mature in one season and usually grow in pairs on a long stalk, sprout soon after they fall from the tree in late summer and autumn. Leaf lobes are shallow, unlike the deeply lobed leaves typical of other members of the white oak group, such as the White Oak (pg. 209) and Bur Oak (pg. 215). Also called Bicolor Oak (from the species name *bicolor*), which refers to its distinctly different upper and lower leaf surfaces. The crooked lower branches tend to hang down, giving it a messy appearance. Does very well as a landscape plant. Its hard, durable wood has been used in making furniture.

fruit

bark

flower

White Oak
Quercus alba

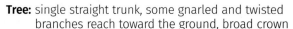

Family: Beech (Fagaceae)

Height: 50–70' (15–21 m)

Tree: single straight trunk, some gnarled and twisted branches reach toward the ground, broad crown

Leaf: lobed, 4–8" (10–20 cm) long, alternately attached, 5–9 rounded lobes, notches deeply cut or shallow and uniform in size and depth, often widest above middle, lacking teeth, bright green above, paler below, leaves often clustered at ends of branches

Bark: light gray, broken into reddish scales

Flower: green catkin, 1–3" (2.5–7.5 cm) long, composed of many tiny flowers, ⅛" (.3 cm) wide

Fruit: green acorn, turns brown, edible, ½–1½" (1–4 cm) long, cap covers the upper third of nut

Fall Color: red-brown

Origin/Age: native; 150–250 years

Habitat: variety of soils, sun

Range: throughout

Stan's Notes: A very important tree in the lumber industry, with its wood used for furniture, flooring, whiskey barrels, crates and much more. Similar to the Bur Oak (pg. 215), which has a single large terminal lobe unlike White Oak's finger-like lobes. Produces edible acorns each fall, with large crops produced every 4–10 years. Like all other acorns, these should be boiled in several changes of water to leach out the bitter and slightly toxic tannins before eating. Acorns are an important food source for turkeys, squirrels, grouse, deer and other wildlife. Susceptible to oak wilt, causing gradual death. Oaks in the white oak group can be treated for oak wilt, while red oak group trees die quickly from the disease.

bark

fruit

Black Oak
Quercus velutina

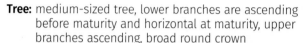

Family: Beech (Fagaceae)

Height: 40–60' (12–18 m)

Tree: medium-sized tree, lower branches are ascending before maturity and horizontal at maturity, upper branches ascending, broad round crown

Leaf: lobed, 4–9" (10–22.5 cm) long, alternately attached, 5–7 lobes, each ending in a pointed tip (bristle-tipped) and separated by deep U-shaped sinuses, shiny green above, yellowish-brown below

Bark: shiny dark gray and smooth texture when young, becoming nearly black with deep reddish cracks

Flower: light yellow catkin, 1–3" (2.5–7.5 cm) long, made up of many tiny flowers, ⅛" (.3 cm) wide

Fruit: green acorn, turns brown at maturity, ¾" (2 cm) long, almost as wide as long, thin black vertical lines on hull, cap covers the upper half of nut

Fall Color: orange-brown

Origin/Age: native; 175–200 years

Habitat: dry sandy soils, steep slopes, sun

Range: throughout, except for the northern tier of counties

Stan's Notes: One of approximately 90 oak species in the U.S. Two oak groups, red and white, with the Black Oak a member of the red oak group. Acorns of the red oak group mature in two seasons, while white oak group acorns mature in one season. Has heavy fruit crops only infrequently. Nuts are bitter due to tannic acid (tannin). Bark also contains tannin, which was used in tanning animal skins. New leaves unfurling in spring are crimson before turning silvery, then dark green. Highly susceptible to oak wilt disease.

bark

flower

fruit

Red Oak
Quercus rubra

Family: Beech (Fagaceae)

Height: 50–70' (15–21 m)

Tree: single straight trunk, broad round crown

Leaf: lobed, 4–9" (10–22.5 cm) long, alternately attached, 7–11 lobes, each lobe ending in several pointed tips (bristle-tipped), with sinuses cutting only halfway to midrib, tufts of hair on veins underneath, dull yellow-green

Bark: dark gray color and smooth texture when young, becoming light gray and deeply furrowed with flat narrow ridges with age

Flower: green catkin, 1–4" (2.5–10 cm) long, composed of many tiny flowers, ⅛" (.3 cm) wide

Fruit: green acorn, turning brown when mature, ½–1" (1–2.5 cm) long, on a short stalk, cap covering the upper quarter of nut

Fall Color: red to brown

Origin/Age: native; 100–150 years

Habitat: moist soils, also does well in dry soils, sun

Range: throughout

Stan's Notes: This is a member of the red oak group. The wood is reddish brown, giving the species name *rubra*, meaning "red," as well as the common name, Red. Differentiated from other red oaks by the leaf sinuses between lobes cutting only halfway to the midrib. The pointed leaves and acorns that mature in two seasons distinguish it from the white oak group, which has rounded leaves and acorns that mature in one season. Its bitter nuts are not popular with wildlife. Succumbs to oak wilt, dying a few weeks after infection. Red Oak wood is used in flooring, furniture and many other products.

bark

flower

fruit

Bur Oak
Quercus macrocarpa

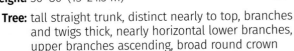

Family: Beech (Fagaceae)

Height: 50–80' (15–24.5 m)

Tree: tall straight trunk, distinct nearly to top, branches and twigs thick, nearly horizontal lower branches, upper branches ascending, broad round crown

Leaf: lobed, 5–12" (12.5–30 cm) in length, alternately attached, 7–9 rounded lobes, last (terminal) lobe often the largest, margin lacking teeth, shiny dark green, leaves clustered near ends of twigs

Bark: dark gray, deeply furrowed, many ridges, scales

Flower: green catkin, 1–3" (2.5–7.5 cm) long, composed of many tiny flowers, ⅛" (.3 cm) wide

Fruit: green acorn, turning brown when mature, sweet and edible, 1–2" (2.5–5 cm) long, cap with hairy edge covering more than the upper half of nut

Fall Color: yellow or brown

Origin/Age: native; 150–250 years

Habitat: deep rich soils, drought and shade tolerant

Range: scattered in the southern half of the state

Stan's Notes: The largest eastern oak, found between prairie and woodland. Thick corky bark allows it to withstand fires. Member of the white oak group (leaves have rounded lobes; acorns mature in one season). Leaves highly variable, but lobes are always rounded, with the terminal lobe the largest. Species name is Latin, with *macro* for "large" and *carpa* for "finger," referring to the large terminal leaf lobe. Heavy fruit crops every three to five years, depending on the weather. Sweet, edible acorns frequently contain Nut Weevil larvae. Often has oak gall, a fleshy, swollen, round deformity caused by a kind of wasp larvae. Also called Blue Oak or Mossycup Oak.

fruit

bark

flower

Boxelder
Acer negundo

Family: Soapberry (Sapindaceae)

Height: 30–50' (9–15 m)

Tree: medium-sized tree, frequently with a divided and crooked trunk, broad irregular crown

Leaf: compound, 4–9" (10–22.5 cm) in length, oppositely attached, made of 3–5 leaflets, each leaflet 2–4" (5–10 cm) in length, often 3-lobed, irregular-toothed margin, pale green

Bark: light gray to tan, becoming deeply furrowed with wavy ridges

Flower: tiny reddish flower, ¼" (.6 cm) wide, growing on a 1–3" (2.5–7.5 cm) long stalk

Fruit: pair of green winged seeds (samara), turning to brown, 1–2" (2.5–5 cm) long

Fall Color: yellow

Origin/Age: native; 50–60 years

Habitat: wet, along streams, lakes and flooded areas, sun

Range: throughout

Stan's Notes: One of the most common trees in the state. Unique species among native maple trees because its leaves are compound. Frequently thought of as a trash tree, but it produces large amounts of seeds that stay on the tree during winter, making a valuable food source for wildlife. If the tree is tapped in spring, it will yield a sap that can be boiled into maple syrup. Since its sugar content is lower than that of the other maples, it takes more sap to make a comparable syrup. These trees are often covered with Boxelder Bugs, harmless beetles whose larvae eat the leaves but cause little damage. Also called Manitoba Maple or Ash-leaved Maple.

bark

fruit

American Bladdernut
Staphylea trifolia

Family: Bladdernut (Staphyleaceae)

Height: 20–25' (6–7.5 m)

Tree: small tree, multiple thin trunks, open crown

Leaf: compound, 6–9" (15–22.5 cm) long, oppositely attached, made of 3–5 oval leaflets, each leaflet 1–3" (2.5–7.5 cm) in length, with 2 pointed, nearly stalkless side leaflets and 1 long-stalked center, fine-toothed margin, dark green above and paler green below

Bark: gray, smooth, often becoming scaly and cracked with age

Fruit: 3-lobed green capsule, turning brown at maturity, 1–2" (2.5–5 cm) long, hanging down, opening at pointed end to release shiny brown round seeds

Fall Color: yellow, turning brown

Origin/Age: native; 30–50 years

Habitat: moist soils, understory of deciduous forests, shade

Range: scattered locations throughout Pennsylvania

Stan's Notes: Often overlooked because it grows in the understory of deciduous forests. Species name *trifolia* refers to the three leaflets (tri-folia) of its compound leaf. Common name refers to its unique inflated green-to-brown bladders. The bladders, which are usually seen in summer and autumn, help to identify this tree. About 50 species of trees in this family with only two native to North America.

bark

flower

fruit

Black Ash

Fraxinus nigra

Family: Olive (Oleaceae)

Height: 40–50' (12–15 m)

Tree: tall slender tree, slender trunk is often leaning or bent, upright branches, open narrow crown

Leaf: compound, 9–17" (22.5–40 cm) long, oppositely attached, made of 7–13 narrow tapered leaflets, each leaflet 3–5" (7.5–12.5 cm) long, with pointed tip, fine-toothed margin, yellowish green, lacking a leaflet stalk (sessile)

Bark: light gray, becoming corky with ridges coming off like scales, soft enough to indent with a fingernail, flaking off when rubbed

Flower: green flower, ⅛" (.3 cm) wide, in loose clusters

Fruit: green winged seed (samara), turning brown when mature, 1–2" (2.5–5 cm) long, in clusters, often remaining on tree well into winter

Fall Color: yellow and brown

Origin/Age: native; 100–125 years

Habitat: wet soils, floodplains, can tolerate standing water for several weeks, shade intolerant

Range: throughout, except for the extreme southern edge

Stan's Notes: Produces a good seed crop every five to seven years, with seeds usually germinating two years after falling from the tree. Sixteen species of ash in the U.S., with six found east of the Rocky Mountains. Black Ash leaflets are not stalked, as are the leaflets of other ash species. Also called Swamp Ash because it often grows at water's edge. Known as Basket Ash or Hoop Ash because the fresh green wood cut into strips was used to weave baskets and make snowshoe frames and canoe ribs.

fruit

bark

White Ash
Fraxinus americana

Family: Olive (Oleaceae)

Height: 40–60' (12–18 m)

Tree: medium-sized tree with single straight trunk and narrow, open round crown

Leaf: compound, 8–12" (20–30 cm) in length, oppositely attached, composed of 7 (occasionally 5–9) oval leaflets, each leaflet 3–5" (7.5–12.5 cm) long, with few teeth or toothless, dark green above, distinctly whiter in color below, on a leaflet stalk (petiolule), ¼–½" (.6–1 cm) long

Bark: greenish gray with many furrows and interlacing diamond-shaped ridges

Fruit: green winged seed (samara), turning brown when mature, 1–2" (2.5–5 cm) long, notched or rounded wing tip, remaining on tree into winter

Fall Color: bronze purple

Origin/Age: native; 150–200 years

Habitat: well-drained soils, upland sites, sun

Range: throughout

Stan's Notes: Very similar to the Green Ash (pg. 225), but tends to grow in sunny, dry, well-drained upland sites. It is by far the most abundant of all 16 ash tree species in the U.S. A fast-growing tree, able to reach a height of up to 10 feet (3 m) in just under 5 years. Produces seeds annually and unusually large masses of seeds every 2–5 years. Young sprouts emerge from stumps or after fire damage. Common name "White" refers to the pale underside of leaves. The straight. narrow-grained wood is used to make baseball bats, snowshoes and hockey sticks.

bark

fruit

Green Ash

Fraxinus pennsylvanica

COMPOUND
OPPOSITE

Family: Olive (Oleaceae)

Height: 50–60' (15–18 m)

Tree: single straight trunk with ascending branches and irregular crown

Leaf: compound, 9–16" (22.5–40 cm) long, oppositely attached, made of 5–9 stalked leaflets, each leaflet 1–2" (2.5–5 cm) long, lacking teeth or with a very fine-toothed margin, on a very short leaflet stalk (petiolule), ⅛" (.3 cm) long

Bark: brown with deep furrows and narrow interlacing ridges, often appearing diamond-shaped

Fruit: green winged seed (samara), turning brown when mature, 1–2" (2.5–5 cm) in length, mostly round-ended, sometimes notched, in clusters, frequently remaining on tree into winter

Fall Color: yellow

Origin/Age: native; 75–100 years

Habitat: wet soils, along streams, lowland forests, shade

Range: throughout

Stan's Notes: This is by far the most widespread of all Pennsylvania ash trees, found throughout the state. Also known as Red Ash because it was once thought that Green Ash and Red Ash trees were separate species. These are now considered one species. Not as water tolerant as the Black Ash (pg. 221), but able to survive with its roots under water for several weeks early in spring. Often has a large unattractive growth (insect gall) at the ends of small branches that persists on the tree throughout the year. The strong white wood is used for baseball bats, skis and snowshoes.

bark

fruit

Common Hoptree

Ptelea trifoliata

Family: Rue (Rutaceae)

Height: 20–25' (6–7.5 m)

Tree: small tree, single straight trunk, branching low to the ground, full round crown

Leaf: compound, 4–7" (10–18 cm) in length, alternately attached, composed of 3 leaflets, each leaflet 2–4" (5–10 cm) long, wider in middle with pointed tip, smooth toothless margin, shiny dark green above, much lighter below with many tiny translucent dots that can be seen when held up to the light, lacking a leaflet stalk (sessile), attaching directly to the central stalk (rachis), which is 3" (7.5 cm) long

Bark: reddish brown, smooth

Fruit: wafer-like flat green disk (samara), turning yellow-brown and papery, up to 1" (2.5 cm) wide, wavy margin, remaining on tree well into winter

Fall Color: yellow

Origin/Age: native; 30–50 years

Habitat: well-drained dry soils, along forest edges, open woodlands, sun to partial shade

Range: scattered locations in southern Pennsylvania

Stan's Notes: A member of the Rue family (Citrus), which explains why its leaves have a strong citrus scent when crushed. Leaflets are often many different sizes. Seeds were used at one time in brewing beer as a substitute for hops, hence the common name. Like many trees with aromatic leaves and bark, it has been used for many ailments for its medicinal properties. Fruit matures late in summer. An easy tree to identify without leaves because its mature yellow-brown fruit remains on the tree. Also called Stinking-ash or Wafer-ash.

bark

flower

fruit

Tree-of-heaven
Ailanthus altissima

Family: Quassia (Simaroubaceae)

Height: 50–70' (15–21 m)

Tree: medium tree, single or multiple crooked trunks, many stout branches, open round crown

Leaf: compound, 12–24" (30–60 cm) long, alternately attached, composed of 13–25 (sometimes more) lance-shaped leaflets, each 3–5" (7.5–12.5 cm) long, toothless except for 2–5 teeth near the base, each tooth accompanied by a tiny gland on underside of leaflet, light green above, paler below

Bark: light brown, smooth, becoming rough with age

Flower: 5-petaled yellowish green flower, ¼" (.6 cm) wide, in large conical clusters, 6–11" (15–28 cm) tall, at the tip of branch

Fruit: green winged seed, turning red at maturity in late summer, flat, 1¼–1½" (3–4 cm) long

Fall Color: yellow

Origin/Age: non-native; 50–75 years

Habitat: wide variety of soils, abandoned fields, old farms

Range: throughout, planted in parks, yards and along streets

Stan's Notes: A fast-growing tree introduced from northern China around 1874. Planted widely as an ornamental due to its hardiness and rapid growth, 5–8 feet (1.5–2.4 m) per year. A very difficult tree to eradicate once established. Male flowers have an unpleasant odor. Flowers bloom after leaves are developed, unlike most native trees. Produces large clusters of winged seeds during late summer, which remain on the tree into winter. Survives well in cities. Presumably the species in Betty Smith's novel *A Tree Grows in Brooklyn*. Native to Asia and northern Australia with about ten *Ailanthus* species.

bark

flower

fruit

European Mountain-ash
Sorbus aucuparia

Family: Rose (Rosaceae)

Height: 15–25' (4.5–7.5 m)

Tree: single trunk, ascending branches, round crown

Leaf: compound, 4–8" (10–20 cm) in length, alternately attached, made of 9–17 oval leaflets, each leaflet 1–2" (2.5–5 cm) long, with fine-toothed margin, dull green above, whitish below, central stalk (rachis) often yellowish

Bark: shiny gray in color, smooth with many horizontal lines (lenticels)

Flower: greenish flower, ¼" (.6 cm) wide, in flat clusters, 3–5" (7.5–12.5 cm) wide

Fruit: orange to red berry-like fruit (pome), ¼" (.6 cm) diameter, in hanging clusters, 3–5" (7.5–12.5 cm) long, each containing 1–2 shiny black seeds

Fall Color: yellow

Origin/Age: non-native, introduced from Europe; 25–50 years

Habitat: dry soils, old fields, abandoned farms, sun

Range: throughout, planted in parks, yards and along roads

Stan's Notes: A member of the Rose family and not a type of ash tree, it is by far the most commonly planted type of mountain-ash tree. Nearly identical to American Mountain-ash (pg. 233), except for the yellowish central stalk. Often a favorite tree of Yellow-bellied Sapsuckers. Spread by birds passing seeds through their digestive tracts unharmed. Latin species name *aucuparia* comes from *avis* and *capere*, meaning "to catch birds," suggesting the fruit was used to bait bird traps. The berry-like fruit provides a good source of food for birds and other wildlife. Containing high amounts of vitamin C, it was once used to cure scurvy. Also called Rowan-tree.

bark

flower

fruit

American Mountain-ash
Sorbus americana

Family: Rose (Rosaceae)

Height: 15–30' (4.5–9 m)

Tree: small tree, often shrub-like with multiple trunks, open round crown

Leaf: compound, 6–9" (15–22.5 cm) in length, alternately attached, with 11–17 lance-shaped leaflets, each leaflet 2–4" (5–10 cm) long, fine sharp teeth, pale green, central stalk (rachis) often reddish

Bark: light gray color, many elongated horizontal lines (lenticels), smooth, becoming very scaly with age

Flower: white-to-cream flower, ¼" (.6 cm) wide, in flat clusters, 3–5" (7.5–12.5 cm) wide

Fruit: bright orange or red berry-like fruit (pome), ¼" (.6 cm) diameter, in hanging clusters, 3–5" (7.5–12.5 cm) wide

Fall Color: yellow

Origin/Age: native; 25–50 years

Habitat: cool moist sites, wetland and forest edges, rocky hillsides, shade

Range: throughout

Stan's Notes: This slow-growing ornamental tree is widely planted in landscapes for its showy flowers and resulting brightly colored fruit. A favorite forage of deer. Often a favorite tree of Yellow-bellied Sapsuckers, which drill rows of horizontal holes and lap up the sap. Birds, especially Cedar Waxwings and Ruffed Grouse, consume the berry-like fruit. Cultivated varieties are often planted to attract birds. A sun-intolerant tree, susceptible to fire blight disease and sunscald. About 75 species of mountain-ash worldwide.

thorn

bark

fruit

Common Prickly-ash
Zanthoxylum americanum

COMPOUND ALTERNATE

Family: Rue (Rutaceae)

Height: 5–15' (1.5–4.5 m)

Tree: small tree, single or multiple trunks, round crown

Leaf: compound, 5–10" (12.5–25 cm) long, alternately attached, made up of 5–11 oval leaflets, each leaflet 1–2" (2.5–5 cm) in length, lacking teeth, citrus scent when crushed, dull green

Bark: gray, smooth with dark gray marks (lenticels) and white blotches, stout sharp thorns on branches

Flower: small, green and inconspicuous on long stalks

Fruit: green berry-like fruit (pome), turning bright red when mature, ¼" (.6 cm) diameter, in clusters, splitting open in fall and releasing the seed, strong orange or lemon-like citrus scent when crushed

Fall Color: yellow

Origin/Age: native; 25–30 years

Habitat: wide variety of soils, along forest edges, sun to partial shade

Range: scattered in south central and southeastern parts of Pennsylvania

Stan's Notes: One of two members of the Rue (sometimes referred to as Citrus) family in the state. This is a small tree with stout sharp thorns. The leaves and especially the berry-like fruit contain zanthoxylin, a citrus-smelling oil. Fruit and inner bark cause numbness in the mouth and have been used for treatment of toothaches, hence its other common name, Toothache Tree. The Greek genus name *Zanthoxylum* means "yellow wood" and describes the wood color. It reproduces from underground roots and forms thick stands along forest edges. The fruit is highly fragrant when it is crushed, smelling like a combination of oranges, lemons and limes.

immature
fruit

bark

fruit

Bitternut Hickory
Carya cordiformis

Family: Walnut (Juglandaceae)

Height: 50–100' (15–30.5 m)

Tree: large tree, sturdy straight trunk, slender upright branches, open round crown

Leaf: compound, 6–10" (15–25 cm) in length, alternately attached, made up of 7–11 narrow leaflets, each leaflet 3–6" (7.5–15 cm) long, with pointed tip and fine-toothed margin, shiny green above, paler in color below, lacks a leaflet stalk (sessile), attaching directly to the central stalk (rachis)

Bark: gray in color with irregular vertical cracks, scaly in appearance but never loose scales

Fruit: nut, too bitter to be edible, round, ¾–1½" (2–4 cm) diameter, with pointed end, 4 ridges extending to point, yellowish hairs covering outer husk

Fall Color: golden yellow

Origin/Age: native; 100–150 years

Habitat: moist soils, lowlands, shade intolerant

Range: throughout

Stan's Notes: The most extensive and northerly of hickories, it is slow growing and shade intolerant. Its large, distinctive yellow buds are diagnostic before the leaves emerge. The wood is used to smoke meat and produces the best flavor of all hickories. Meat of the nuts is very bitter and unpalatable to humans and much wildlife, hence the common name. Oil extracted from the nuts was used for lamp fuel. Sometimes called Bitter Pecan because it is closely related to the familiar pecan. Also called Swamp Hickory due to its preference for wet or loamy soils. Hickories are found naturally in eastern North America and Asia. About 12 species are native to North America.

fruit

bark

Red Hickory
Carya glabra

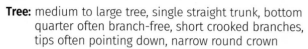

Family: Walnut (Juglandaceae)

Height: 60–80' (18–24.5 m)

Tree: medium to large tree, single straight trunk, bottom quarter often branch-free, short crooked branches, tips often pointing down, narrow round crown

Leaf: compound, 6–10" (15–25 cm) in length, alternately attached, composed of 5 (rarely 7) lance-shaped leaflets, each leaflet 3–6" (7.5–15 cm) long, wider in middle, fine-toothed margin, midrib covered with hair, yellowish green above and paler below, terminal and first pair of leaflets larger than leaflets closest to stem, nearly stalkless

Bark: gray, becoming very shaggy with age, often peeling into long, thin, loosely attached strips

Fruit: pear-shaped to round green nuts, turning brown when mature, 1–2" (2.5–5 cm) diameter, in small clusters, thin shell splitting open into 4 sections, inedible nut

Fall Color: yellow

Origin/Age: native; 150–200 years

Habitat: well-drained dry soils, hillsides, dry ridges, sun

Range: southern half of the state

Stan's Notes: One of the most common hickories in the eastern U.S. Also called Pignut Hickory because the nuts were considered fit only for hogs. Flavor is actually highly variable—some nuts are sweet, others are bitter. Was called Broom Hickory because broom handles were made from saplings. Wood was once used in wagon wheels. Usually in mixed stands of deciduous trees in upland sites, unlike Shellbark Hickory (pg. 245), which grows in moist to wet sites.

bark

fruit

twig pith

Shagbark Hickory
Carya ovata

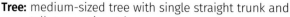

Family: Walnut (Juglandaceae)

Height: 40–60' (12–18 m)

Tree: medium-sized tree with single straight trunk and tall, narrow irregular crown

Leaf: compound, 8–14" (20–36 cm) in length, alternately attached, with 5 (rarely 7) pointed leaflets, each leaflet 3–4" (7.5–10 cm) long, widest at the middle, upper 3 leaflets larger than lower 2, fine-toothed margin, yellowish green, lacks leaflet stalk (sessile), attaching directly to central stalk (rachis)

Bark: gray in color, long smooth vertical strips curling at each end, giving it a shaggy appearance

Fruit: green nut, turning brown at maturity, inner kernel sweet, edible, round to oval, 1–1½" (2.5–4 cm) in diameter, single or in pairs, thick 4-ribbed husk

Fall Color: yellow

Origin/Age: native; 150–200 years

Habitat: rich moist soils, hillsides, slopes, bottomlands, sun

Range: throughout

Stan's Notes: Also called Upland Hickory. Often found on hillsides that have rich moist soils, growing branch-free for three-quarters of its height. Its common name comes from the large scaly or "shaggy" bark. Also known as Shell-bark or Seal-bark Hickory. Hickories are divided into two groups: true hickories, which include Shagbark, and pecan hickories, which include Bitternut Hickory (pg. 237). Shagbark nuts are eaten by wildlife and people. Its extremely hard wood is used to make tool handles, skis and wagon wheels. Unlike the twigs of walnut trees, which have a light brown pith, twigs of this species have a white pith (see inset).

241

bark

immature fruit

fruit

Mockernut Hickory

Carya tomentosa

COMPOUND ALTERNATE

Family: Walnut (Juglandaceae)

Height: 40–80' (12–24.5 m)

Tree: medium to large tree, usually has a straight trunk, branchless to halfway up, narrow round crown

Leaf: compound, 8–20" (20–50 cm) in length, alternately attached, made of 7–9 leaflets, each leaflet 2–8" (5–20 cm) in length, elliptical, pointed at tip, round at base, wider middle, fine-toothed, shiny dark green above, paler and hairy below, leaflet nearly stalkless

Bark: gray to light brown with narrow forked ridges

Fruit: thick-shelled green nut, 1½–2" (4–5 cm) diameter, turning brown when mature, inner kernel edible, tan to light brown, ¾–1" (2–2.5 cm) wide, 4-sided

Fall Color: yellow

Origin/Age: native; 300–500 years

Habitat: moist upland sites, frequently with oaks and other hickories on ridges and hillsides, sun

Range: southern half of the state

Stan's Notes: A straight-growing hickory, common in the eastern half of the U.S. Wood has been valued for its strength for furniture; also used for smoking meat, such as ham. Species name comes from the Latin word *tomentum*, meaning "covered with dense short hairs," referring to the underside of leaves and helping identify the species. Also called White Hickory due to the light color of the wood. The common name "Mockernut" comes from the large thick-shelled fruit with very small kernels of meat inside. Its nuts are an important wildlife food; many birds and animals eat or store them for winter. Produces nuts after 20 years, but the prime nut-bearing age is from 50 to 150 years old. Stout, hairy twigs, often reddish brown.

bark

fruit

Shellbark Hickory
Carya laciniosa

COMPOUND
ALTERNATE

Family: Walnut (Juglandaceae)

Height: 70–90' (21–27.5 m)

Tree: large tree, single straight trunk, often branch-free on the bottom half, narrow round crown

Leaf: compound, 12–20" (30–50 cm) in length, alternately attached, composed of 7 (rarely 9) lance-shaped leaflets, each leaflet 2–8" (5–20 cm) in length, fine-toothed margin, shiny green above, paler below and covered with soft hair

Bark: gray, becoming rough and shaggy with age, often peeling into long, thin, loosely attached strips

Fruit: thick-shelled green fruit, turning dark brown at maturity, edible, round, 2–3" (5–7.5 cm) diameter, husk splits open into 4 sections, releasing a nearly round nut

Fall Color: yellow

Origin/Age: native; 150–200 years

Habitat: moist to wet soils, floodplains, sun to part shade

Range: southwestern quarter and in scattered locations in south central and southeastern Pennsylvania

Stan's Notes: The Shellbark's large leaves, extremely large fruit and orange twigs make it one of the easiest hickories to identify. Orange twigs are distinctive, but not always obvious. Similar scaly bark as Shagbark Hickory (pg. 241), but the Shellbark has more leaflets and larger fruit. The Latin species name *laciniosa* means "with folds" and refers to the characteristic shaggy bark. Also known as Big Shagbark Hickory or Kingnut, referring to the oversized fruit. Fruit is edible and eaten by wildlife. Central stalk (rachis) remains on the tree after the leaflets fall each autumn.

flower

bark

fruit

Poison Sumac

Toxicodendron vernix

Family: Cashew (Anacardiaceae)

Height: 5–20' (1.5–6 m)

Tree: small tree, single trunk, few horizontal branches, open irregular crown

Leaf: compound, 6–12" (15–30 cm) in length, alternately attached, composed of 7–13 lance-shaped leaflets, each leaflet 1½–3" (4–7.5 cm) long, lacking teeth, smooth to touch, dark green above, pale white in color below, leaflet stalk (petiolule) and central stalk (rachis) often reddish

Bark: light gray, smooth

Flower: green flower, ¼" (.6 cm) wide, in clusters, 2–4" (5–10 cm) wide, on open branches

Fruit: green-to-glossy-white berry-like fruit (drupe), ¼" (.6 cm) diameter, remaining on tree into winter

Fall Color: yellow to red

Origin/Age: native; 50–75 years

Habitat: wet soils, bogs, sun

Range: scattered throughout

Stan's Notes: The only sumac of the three species growing in the state that is poisonous. A rare tree found only in open swamps and bogs. Few if any encounter this tree due to its remote wet habitat and the rarity of the species. The oils of Poison Sumac are toxic, causing severe skin rash. If burned, the smoke can cause severe breathing difficulties and irritated eyes and skin. Closely related to other sumacs in Pennsylvania (pgs. 249 and 251) and to Poison Ivy.

flower

bark

fruit

Smooth Sumac

Rhus glabra

Family: Cashew (Anacardiaceae)

Height: 10–20' (3–6 m)

Tree: single or multiple trunks, closed flat-topped crown

Leaf: compound, 12–24" (30–60 cm) in length, alternately attached, with 11–31 leaflets, each leaflet 2–4" (5–10 cm) long, toothed margin, dark green above, some red hairs below, lacking a leaflet stalk (sessile), attaching directly to central stalk (rachis)

Bark: brown, smooth, rarely any furrows or scales

Flower: green flower, ¼" (.6 cm) diameter, in open, wide upright clusters, 4–8" (10–20 cm) tall

Fruit: red berry-like fruit (drupe), ⅛" (.3 cm) diameter, in cone-shaped clusters, 4–8" (10–20 cm) long

Fall Color: red

Origin/Age: native; 25–50 years

Habitat: dry or poor soils, forest edges, sun

Range: throughout

Stan's Notes: One of the first trees to turn colors in fall. Over 100 species of sumac, most occurring in southern Africa. Closely related to the Staghorn Sumac (pg. 251), which has hairy central stalks and leaves. While Smooth Sumac is not as common as Staghorn, they hybridize where occurring together. Fast growing, reproducing by underground roots that send up new trunks. Often forms a dense stand. Once established, it is often hard to eradicate. Has been planted to stabilize slopes from erosion. Male and female flowers are on separate trees (dioecious), so not all sumacs produce the attractive clusters of red fruit. Animals and birds eat the berry-like fruit. Ripe sumac fruit is used to make a lemonade-like drink.

flower

bark

fruit

Staghorn Sumac
Rhus typhina

Family: Cashew (Anacardiaceae)

Height: 10–20' (3–6 m)

Tree: single or multiple trunks, closed flat-topped crown

Leaf: compound, 12–24" (30–60 cm) in length, alternately attached, composed of 11–31 leaflets, each leaflet 2–4" (5–10 cm) long, with toothed margin, dark green above, red hairs below, lacking a leaflet stalk (sessile), attaching directly to the central stalk (rachis), which is often hairy and reddish

Bark: brown, smooth, rarely any furrows or scales

Flower: green flower, ¼" (.6 cm) diameter, in tight cone-shaped clusters, 4–8" (10–20 cm) long

Fruit: fuzzy red berry-like fruit (drupe), ⅛" (.3 cm) wide, in thin cone-shaped clusters, 4–8" (10–20 cm) long

Fall Color: red to maroon

Origin/Age: native; 25–50 years

Habitat: dry or poor soils, forest edges, sun

Range: throughout, along roads, in parks and yards

Stan's Notes: One of the first trees to turn colors each autumn. Its fuzzy central stalks resemble the velvety antlers of a deer, hence the common name. Closely related to Smooth Sumac (pg. 249), which lacks hairy leaves and central stalks. While Staghorn Sumac is by far more common than the Smooth, they will hybridize where they occur together. Reproduces by underground roots that send up new trunks. Fast growing, often forming a dense stand that extends up to several hundred feet in each direction. Often planted along highways for its rapid growth, thus stabilizing the soil quickly. Animals and birds eat the red berry-like fruit. "Sumac-ade" can be made from ripe sumac fruit.

thorn

bark

flower

fruit

Black Locust

Robinia pseudoacacia

Family: Pea or Bean (Fabaceae)

Height: 30–50' (9–15 m)

Tree: medium-sized tree, often crooked trunk, upright spreading branches, open irregular crown

Leaf: compound, 7–14" (18–36 cm) in length, alternately attached, composed of 7–19 oval to round leaflets, each leaflet 1–2" (2.5–5 cm) in length, lacks teeth, yellowish green

Bark: dark brown color and smooth texture, becoming furrowed and scaly with age, opposite pairs of stout thorns (see inset), especially on young trees

Flower: pea-like white flower with yellow center, ½–1" (1–2.5 cm) wide, in hanging clusters, 2–4" (5–10 cm) long, appearing soon after leaves develop, fragrant

Fruit: green pod, turning brown when mature, flat, 2–4" (5–10 cm) long, containing 4–8 seeds

Fall Color: yellow

Origin/Age: native; 75–100 years

Habitat: adapts to almost any type of soil, moist woods, shade intolerant

Range: southwestern Pennsylvania, planted throughout along roads, in parks and around homes

Stan's Notes: Fifteen species of locust trees and shrubs, all native to North America. Wood is so strong that in the 1800s the British credited success of the U.S. naval fleet in the War of 1812 to Black Locust lumber, used to build ships. Roots combine with bacteria, fixing atmospheric nitrogen into the soil. Spreads rapidly by root suckering. Branches and twigs have stout thorns in pairs at bases of leafstalks. Susceptible to Locust Borer beetles, which bore into trunks.

twig pith

bark

flower

fruit

Black Walnut
Juglans nigra

Family: Walnut (Juglandaceae)

Height: 50–75' (15–23 m)

Tree: straight trunk, open round crown

Leaf: compound, 12–24" (30–60 cm) in length, alternately attached, with 15–23 stalkless leaflets (sessile), each leaflet 3–4" (7.5–10 cm) long, with pointed tip, last (terminal) leaflet often smaller or absent, middle leaflets larger than on either end, fine-toothed margin, yellowish green and smooth above, slightly lighter and hairy below

Bark: brown to black, becoming darker with age, deep pits and flat scaly ridges

Flower: catkin, 2–4" (5–10 cm) long, composed of many tiny green flowers, ¼" (.6 cm) wide

Fruit: fleshy green fruit, round, 1–2" (2.5–5 cm) wide, in clusters, aromatic green husk surrounding a hard dark nut that matures in autumn, nutmeat sweet and edible

Fall Color: yellowish green

Origin/Age: native; 150–175 years

Habitat: well-drained rich soils, shade intolerant

Range: scattered in the southern half of the state

Stan's Notes: One of six species of walnut native to North America. Valued for its wood, which doesn't shrink or warp and is used to build furniture and cabinets. An important food source for wildlife. Fruit husks contain a substance that stains skin and were used by pioneers to dye clothing light brown. Twigs have a light brown chambered pith (see inset), unlike dark brown pith of Butternut (pg. 257). Fallen leaves and roots produce juglone, a natural herbicide.

immature
fruit

bark

fruit

twig pith

Butternut
Juglans cinerea

COMPOUND ALTERNATE

Family: Walnut (Juglandaceae)

Height: 40–60' (12–18 m)

Tree: medium-sized tree, usually a divided trunk, open, broad and often flat crown

Leaf: compound, 15–25" (37.5–63 cm) long, alternately attached, made of 11–17 stalkless leaflets (sessile), each leaflet 2–4" (5–10 cm) in length, fine-toothed margin, last (terminal) leaflet usually present and same size as the lateral leaflets, and progressing smaller toward the leaf base, stout hairs covering each leaflet and the central stalk (rachis)

Bark: light gray with broad flat ridges

Flower: catkin, 1–2" (2.5–5 cm) long, composed of many tiny green flowers, ¼" (.6 cm) wide

Fruit: nut, edible, oval, 2–3" (5–7.5 cm) long, in clusters, sticky green husk, turning brown

Fall Color: yellow

Origin/Age: native; 80–100 years

Habitat: wide variety of soils, often on slopes having well-drained rich soils, sun

Range: throughout

Stan's Notes: A short-lived tree, also called White Walnut. Its very hard, strong wood is much sought by woodworkers. Twigs are stout with a dark brown chambered pith (see inset). The sap can be boiled to make syrup. Yellow dye can be extracted from the husks and used to color fabrics. Common name comes from its butter-like oil, which American Indians once extracted from the nuts. Butternut canker, caused by a fungus, has killed many of these trees.

257

bark

flower

fruit

Honey Locust
Gleditsia triacanthos

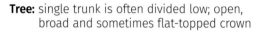

Family: Pea or Bean (Fabaceae)

Height: 40–60' (12–18 m)

Tree: single trunk is often divided low; open, broad and sometimes flat-topped crown

Leaf: twice compound, 12–24" (30–60 cm) long, alternately attached, composed of 14–30 oval, feathery leaflets, each leaflet 1" (2.5 cm) in length, fine-toothed margin, dark green above and yellow-green below

Bark: reddish brown covered with gray horizontal lines (lenticels), often cracking and peeling, frequently thorny on trunk and branches

Flower: small green catkin, 1–2" (2.5–5 cm) long

Fruit: large pea-like purplish-brown pod, flat, twisted, 6–16" (15–40 cm) long, 12–14 oval seeds per pod

Fall Color: yellow

Origin/Age: native; 100–125 years

Habitat: moist or rich soils, sun

Range: southwestern and central Pennsylvania, planted in parks, yards and along roads

Stan's Notes: Also called Thorny Locust because it has large thorns on the trunk and branches and twigs are zigzagged with thorns at joints. Between seeds in seedpods is a sweet yellowish substance, hence "Honey" in the common name. Seedpods are large, obvious and eaten by wildlife. Should not be pruned in wet weather since it opens the tree to infection by Nectria canker. Thornless and seedless varieties are widely planted in parks and yards and along roads. Most varieties planted in landscaping lack thorns and fruit. Largest of the two species of *Gleditsia* native to North America.

bark

fruit

Kentucky Coffeetree
Gymnocladus dioicus

Family: Pea or Bean (Fabaceae)

Height: 40–60' (12–18 m)

Tree: single trunk, can be divided low, many crooked branches, open round crown

Leaf: twice compound, 12–36" (30–90 cm) long, alternately attached, made of many (up to 70) leaflets, each leaflet 2" (5 cm) long, lacks teeth, blue-green

Bark: brown and smooth when young, thin scales with edges curling out, breaking with age into plates

Flower: 5-petaled white flower, ¼–½" (.6–1 cm) diameter, on a single long stalk, in open clusters, 1–3" (2.5–7.5 cm) wide

Fruit: leathery green pod, turning reddish brown when mature, 4–10" (10–25 cm) in length, often covered with a whitish powder, containing 6–9 large seeds

Fall Color: yellow

Origin/Age: native; 50–75 years

Habitat: deep rich soils, sun

Range: throughout, planted in urban sites, farms, parks, along roads

Stan's Notes: The leaves of this species are among the last to appear in spring and first to turn in fall. Common name comes from its coffee bean-like seeds in seedpods. The seedpods contain many dark seeds surrounded by a yellowish pulp that becomes soapy when wet. The bitter seeds are seldom eaten by wildlife and remain viable for several years. New trees sprout from roots of parent trees, often forming small colonies. The genus name *Gymnocladus* comes from the Greek for "naked branch," which is how the branches appear.

cross section

bark

flower

fruit

Yellow Buckeye
Aesculus flava

PALMATE
COMPOUND
OPPOSITE

Family: Soapberry (Sapindaceae)

Height: 50–70' (15–21 m)

Tree: single or multiple trunks with spreading branches, round crown

Leaf: palmate compound, 5–14" (12.5–36 cm) in length, oppositely attached, composed of 5–7 leaflets, each leaflet 4–6" (10–15 cm) long, radiates from a central point, evenly fine-toothed, yellowish green above, lighter below, often hairy, short stalk (petiole)

Bark: brown to gray, thin

Flower: green-to-tan tubular flower, 1¼" (3 cm) long, in triangular clusters, 4–6" (10–15 cm) tall

Fruit: smooth, leathery, 3-parted, light brown capsule, round, 2–3" (5–7.5 cm) wide, thin husk, contains 1–2 poisonous seeds

Fall Color: yellow to orange

Origin/Age: native; 60–80 years

Habitat: deep moist soils, river and mountain valleys

Range: extreme southwestern corner of the state, planted in parks, yards and along streets

Stan's Notes: The largest and tallest buckeye. Never in pure stands in Pennsylvania and confined to rich bottomlands. Commercial species sold as a shade tree. Looks like Ohio Buckeye (pg. 265) but is larger with smooth fruit capsules and hairier leaf undersides. Its soft wood is rated at the bottom of 35 leading timbers of the U.S. Pale yellow wood is used for pulpwood, artificial limbs and interior finishes for homes. Seeds and young shoots are poisonous and make livestock ill. Occurs in a wide variety of habitats up to 6,000 feet (1,830 m). Common tree of the Great Smoky Mountains. Also called Sweet Buckeye, Big Buckeye or Large Buckeye.

fruit

bark

flower

Ohio Buckeye
Aesculus glabra

Family: Soapberry (Sapindaceae)

Height: 20–40' (6–12 m)

Tree: single trunk, broad round crown with flat top

Leaf: palmate compound, 5–15" (12.5–37.5 cm) long, oppositely attached, composed of 5 leaflets, each leaflet 3–5" (7.5–12.5 cm) long, radiating from a central point, with fine irregular teeth, yellowish green above, pale and hairy below, sessile

Bark: brown with scaly patches, rough shallow furrows

Flower: green flower, ½" (1 cm) wide, growing upright in triangular clusters, 5–7" (12.5–18 cm) long, foul odor when crushed

Fruit: light brown spiny capsule, round, 1–2" (2.5–5 cm) wide, contains 1–2 shiny brown poisonous seeds

Fall Color: yellow to orange

Origin/Age: native; 100–125 years

Habitat: moist soils, river bottoms, sun to partial shade

Range: extreme southwestern quarter of the state, planted in parks, yards and along streets

Stan's Notes: Also known as Fetid Buckeye or Stinking Buckeye, referring to the foul odor of the flowers and most other parts of the tree when crushed. Grows naturally in moist areas. Planted as a landscape tree in dry upland areas for its attractive autumn foliage. Its large poisonous seeds are avoided by wildlife. A unique palmate leaf, the leaflets lack their own leafstalks, all rising instead from a central stalk. An extract from the bark was once used as a stimulant for the cerebrospinal system. Once thought a buckeye seed carried in the pocket would ward off rheumatism. Warty spines on its fruit capsules help distinguish it from Yellow Buckeye (pg. 263).

bark

flower

fruit

Horse-chestnut
Aesculus hippocastanum

Family: Soapberry (Sapindaceae)

Height: 40–60' (12–18 m)

Tree: medium-sized tree with single trunk that is often divided low, spreading round crown

Leaf: palmate compound, 5–10" (12.5–25 cm) in length, oppositely attached, composed of 5–9 (usually 7) leaflets, each leaflet 4–10" (10–25 cm) long, widest above the middle, radiating from a central point, with a sharp-toothed margin, hairy below when young, lacking hairs when mature

Bark: dark brown, many furrows and scales, inner bark orange-brown

Flower: white flower with yellow or orange center, ½–1" (1–2.5 cm) wide, upright in spike clusters, 8–12" (20–30 cm) long

Fruit: thick-walled leathery green capsule, rounded, 2" (5 cm) diameter, covered with pointed spines, in hanging clusters, splits in 3 sections, contains 1–3 smooth, non-edible, shiny chestnut-brown seeds

Fall Color: yellow

Origin/Age: non-native, introduced to the U.S. from Europe; 75–100 years

Habitat: wide variety of soils, sun

Range: throughout, planted in parks and yards

Stan's Notes: Closely related to Ohio Buckeye (pg. 265). A remedy made from the seeds was used to treat cough in horses, hence its species and common names, *hippos* ("horse") and *kastanon* ("chestnut"). The chemical esculin has been extracted from its leaves and bark for use in skin protectants. Also called Chestnut.

GLOSSARY

Acorn: A nut, typically of oak trees, as in the White Oak. See *nut* and *fruit*.

Aggregate fruit: A fruit composed of multiple tiny berries, such as a mulberry, raspberry or blackberry. See *fruit*.

Alternate: A type of leaf attachment in which the leaves are singly and alternately attached along a stalk, as in Quaking Aspen.

Arcuate: Curved in form, like a bow, as in veins of Alternate-leaf Dogwood leaves.

Asymmetrical leaf base: A base of a leaf with lobes unequal in size or shape, as in elms. See *leaf base*.

Berry: A fleshy fruit with several seeds within, such as European Buckthorn. See *fruit*.

Bract: A petal-like structure on a flower, as in Blue Beech.

Branch: The smaller, thinner, woody parts of a tree, usually bearing the leaves and flowers.

Bristle-tipped: A type of leaf lobe ending in a projection, usually a sharply pointed tip, as in Red Oak.

Buttress: A wide or flared base of a tree trunk that helps to hold the tree upright in unstable soils, as in Bald Cypress.

Capsule: A dry fruit that opens along several seams to release the seeds within, as in Ohio Buckeye. See *pod*.

Catkin: A scaly cluster of usually same sex flowers, as in Bigtooth Aspen or any willow.

Chambered pith: The central soft part of a twig that is broken into spaced sections. See *pith*.

Clasping: A type of leaf attachment without a leafstalk in which the leaf base grasps the main stalk, partly surrounding the stalk at the point of attachment.

Clustered needles: A group of needles emanating from a central point, usually within a papery sheath, as in pine trees.

Compound leaf: A single leaf composed of at least 2 but usually not more than 20 leaflets growing along a single leafstalk, as in Smooth Sumac.

Cone: A cluster of woody scales encasing multiple nutlets or seeds and growing on a central stalk, as in coniferous trees.

Cone scale: An individual overlapping projection, often woody, on a cone, as in Austrian Pine.

Conifer: A type of tree that usually does not shed all of its leaves each autumn, such as pine or spruce.

Crooked: Off-center or bent in form, not straight, as in a Black Locust trunk.

Deciduous: A type of tree that usually sheds all of its leaves each autumn, such as White Oak or Sugar Maple.

Dioecious: A type of tree that has male and female flowers on separate trees of the same species, as in Quaking Aspen. See *monoecious*.

Disk: A flattened, disk-like fruit that contains a seed, as in the American Elm. See *samara*.

Double-toothed margin: A jagged or serrated leaf edge that is composed of two types of teeth, usually one small and one large, as in Siberian Elm.

Drupe: A fleshy fruit that usually has a single seed, such as a cherry. See *fruit*.

Flower: To bloom, or produce a flower or flowers as a means of reproduction, as in deciduous trees.

Fruit: A ripened ovary or reproductive structure that contains one or more seeds, such as a nut or berry.

Furrowed: Having longitudinal channels or grooves, as in Swamp White Oak bark.

Gall: An abnormal growth of plant tissue that is usually caused by insects, microorganisms or injury.

Gland: An organ or structure that secretes a substance, as in Nannyberry leafstalks.

Intolerant: Won't thrive in a particular condition, such as shade.

Lance-shaped: Long, narrow and pointed in form, like a spear-head, as in Weeping Willow leaves.

Leaf base: The area where a leafstalk attaches to the leaf.

Leaflet: One of the two or more leaf-like parts of a compound leaf, as in White Ash.

Leafstalk: The stalk of a leaf, extending from the leaf base to the branch. See *petiole*.

Lenticel: A small growth, usually on bark, that allows air into the interior of a tree, as in Paper Birch.

Lobed leaf: A single leaf with at least one indentation (sinus or notch) along an edge that does not reach the center or base of the leaf, as in oaks or maples.

Margin: The edge of a leaf.

Midrib: The central vein of a leaf, often more pronounced and larger in size than other veins, as in Black Cherry.

Monoecious: A type of tree that has male and female flowers on the same tree, as in Paper Birch. See *dioecious*.

Naturalized: Not originally native, growing and reproducing in the wild freely now, such as Russian-olive.

Needle: A long, usually thin, evergreen leaf of a conifer.

Notch: A small indentation along the margin of a leaf, as in Red Maple.

Nut: A large fruit encased by hard walls, usually containing one seed, such as an acorn. See *fruit*.

Nutlet: A small or diminutive nut or seed, usually contained in a cone or cone-like seed catkin, as in Red Pine or Paper Birch. See *fruit*.

Opposite: A type of leaf attachment in which leaves are situated directly across from each other on a stalk, as in Sugar Maple.

Ovate: Shaped like an egg, as in Austrian Pine cones.

Palmate compound leaf: A single leaf that is composed of three or more leaflets emanating from a common central point at the end of the leafstalk, as in Ohio Buckeye.

Petiole: The stalk of a leaf. See *leafstalk*.

Petiolule: The stalk of a leaflet in a compound leaf.

Pitch pocket: A raised blister that contains a thick resinous sap, as in Balsam Fir bark.

Pith: The central soft part of a twig in a young branch, turning to hard wood when mature.

Pod: A dry fruit that contains many seeds and opens at maturity, as in Kentucky Coffeetree. See *capsule*.

Pollination: The transfer of pollen from the male anther to the female stigma, usually resulting in the production of seeds.

Pome: A fleshy fruit with several chambers that contain many seeds, such as an apple. See *fruit*.

Rachis: The central or main stalk of a compound leaf, as in the European Mountain-ash.

Samara: A winged fruit that contains a seed, as in maples, ashes or elms. See *disk* and *fruit*.

Seed catkin: A small cone-like structure that contains nutlets or seeds, as in birches.

Sessile: Lacking a stalk and attaching directly at the base, as in Common Hoptree leaflets.

Simple leaf: A single leaf with an undivided or unlobed edge, as in American Elm.

Sinus: The recess or space in between two lobes of a leaf, as in the Red Oak.

Spine: A stiff, usually short, sharply pointed woody outgrowth from a branch or cone, as in Pitch Pine cones. See *thorn*.

Stalk: A thin structure that attaches a leaf, flower or fruit to a twig or branch.

Stipule: An appendage at the base of a stalk, usually small and in pairs, with one stipule on each side of stalk, as in Nannyberry.

Sucker: A secondary shoot produced from the base or roots of a tree that gives rise to a new plant, as in Quaking Aspen.

Tannin: A bitter-tasting chemical found within acorns and other parts of a tree, as in oaks.

Taproot: The primary, vertically descending root of a mature tree.

Terminal: Growing at the end of a stalk or branch.

Thorn: A stiff, usually long and sharply pointed woody outgrowth from a branch or trunk, as in Canada Plum. See *spine*.

Tolerant: Will thrive in a particular condition, such as shade.

Understory: The small trees and other plants that grow under a canopy of larger trees; the shady habitat beneath a forest canopy.

Whorl: A ring of three or more leaves, stalks or branches arising from a common point, as in Red Pine or Northern Catalpa.

Winged: Having thin appendages, attached to a seed, branch or twig, as in maple seeds.

Woody: Composed of wood, as in trees or cones. See *cone scale*.

CHECKLIST/INDEX *Use the boxes to check trees you've seen.*

MORE FOR PENNSYLVANIA BY STAN TEKIELA

Identification Guides

Birds of Prey of the Northeast Field Guide

Birds of the Northeast

Stan Tekiela's Birding for Beginners: Northeast

Birds of Pennsylvania Field Guide

Nature Books

Bird Trivia

Start Mushrooming

A Year in Nature with Stan Tekiela

Children's Books: Adventure Board Book Series

Floppers & Loppers

Paws & Claws

Peepers & Peekers

Snouts & Sniffers

Children's Books

C is for Cardinal

Can You Count the Critters?

Critter Litter

Children's Books: Wildlife Picture Books

Baby Bear Discovers the World

The Cutest Critter

Do Beavers Need Blankets?

Hidden Critters

Jump, Little Wood Ducks

Some Babies Are Wild

Super Animal Powers

What Eats That?

Whose Baby Butt?

Whose Butt?

Whose House Is That?

Whose Track Is That?

ABOUT THE AUTHOR

Naturalist, wildlife photographer and writer **Stan Tekiela** is the originator of the popular state-specific field guide series that includes *Birds of Pennsylvania Field Guide*. Stan has authored more than 190 educational books, including field guides, quick guides, nature books, children's books, playing cards and more, presenting many species of animals and plants.

With a Bachelor of Science degree in Natural History from the University of Minnesota and as an active professional naturalist for more than 30 years, Stan studies and photographs wildlife throughout the United States and Canada. He has received various national and regional awards for his books and photographs. Also a well-known columnist and radio personality, he has a syndicated column that appears in more than 25 newspapers, and his wildlife programs are broadcast on a number of Midwest radio stations. Stan can be followed on Facebook and Twitter. He can be contacted via www.naturesmart.com.